4th International Conference on Design, Analysis, Manufacturing and Simulation (ICDAMS 2020)

Held online due to COVID-19

IOP Conference Series: Materials Science and Engineering
Volume 992

Chennai, India
16 - 17 September 2020

ISBN: 978-1-7138-3181-5
ISSN: 1757-8981

Printed from e-media with permission by:

Curran Associates, Inc.
57 Morehouse Lane
Red Hook, NY 12571

proceedings.com

Some format issues inherent in the e-media version may also appear in this print version.

Institute of Physics
Dirac House, Temple Back
Bristol BS1 6BE UK

Phone: 44 1 17 929 7481
Fax: 44 1 17 920 0979

techtracking@iop.org

Additional copies of this publication are available from:

Curran Associates, Inc.
57 Morehouse Lane
Red Hook, NY 12571 USA
Phone: 845-758-0400
Fax: 845-758-2633
Email: curran@proceedings.com
Web: www.proceedings.com

TABLE OF CONTENTS

IOP Conf. Series: Materials Science and Engineering **992** (2020) 011001 doi:10.1088/1757-899X/992/1/011001

PREFACE

As the Convener of this International Conference, I am delighted to extend you all a very warm welcome on behalf of the ICDAMS 2020 Organizing Team. It is with great pleasure that I introduce the proceedings of the 4th International Conference on Design, Analysis, Manufacturing and Simulation (ICDAMS 2020) which was held online at Saveetha School of Engineering, Saveetha Institute of Medical and Technical Sciences, Chennai, Tamil Nadu, India during 16-17 September 2020 organised by the Departments of Mechanical, Automobile and Agricultural Engineering. The main goal of organizing this conference is to provide a common platform to bring together the academic and industrial communities to address new challenges, share solutions, and discuss future research directions in the field of Engineering in general and in the areas of Design, Manufacturing, and Simulation in particular. The conference program is consisted of keynote address and oral presentations given by renowned scholars, students and faculty members.

ICDAMS 2020 offered the opportunity to all the global researchers to share their view through digital platform without any restrictions due to COVID-19. To enrich the quality of program, some selected papers from 4th International Conference on Design, Analysis, Manufacturing and Simulation were invited to be presented. Distinguished researchers from India, Saudi Arabia and London were invited to deliver keynote address regarding the latest research discoveries in their respective areas of expertise. Authors made oral presentations and each presentation was confined in 15 minutes, following by question and answer. I hope that the participants will get ample opportunities to widen their knowledge and network. The conference was managed with Cisco Webex and Goggle Meet, to discuss author's research work. All accepted papers were subject to peer-review by conference committee and invited reviewers. The papers were selected based on their quality and their relevance to the conference.

I would like to thank the reviewers for their valuable time and advice which helped in improving the quality of the papers selected for presentation at the conference and for publication in the proceedings. Finally I wish to register my heartiest thanks to all the ICDAMS 2020 organizing committee members for extending their valuable time & effort and to all the International and National advisory members, Keynote speakers and authors for their sparkling efforts and their belief in the success of ICDAMS 2020.

Dr.T.SHAAFI

CONVENER – ICDAMS 2020

IOP Conf. Series: Materials Science and Engineering **992** (2020) 011001 doi:10.1088/1757-899X/992/1/011001

CONFERENCE ORGANIZATION

Chief Patron	**Dr.N.M.Veeraiyan**, Chancellor, Saveetha Institute of Medical and Technical Sciences
Executive Patron	**Dr.Deepak Nallaswamy** Director of Academics, Saveetha Institute of Medical and Technical Sciences
	Mrs. Ramya Deepak, Director, Saveetha School of Engineering
Patrons	**Prof. Dr.Rakesh Kumar Sharma**, Vice Chancellor, Saveetha Institute of Medical and Technical Sciences
	Dr.V.Thiyagarajan, Registrar Saveetha Institute of Medical and Technical Sciences
Co- Patron	**Dr.D.Dhanasekaran,** Principal Saveetha School of Engineering
Conference Chair	**Dr.A.Manju,** Professor & Head Department of Mechanical Engineering Saveetha School of Engineering
Convener	**Dr.T.Shaafi,** Professor & Head, Department of Agricultural Engineering, Saveetha School of Engineering

IOP Conf. Series: Materials Science and Engineering **992** (2020) 011001 doi:10.1088/1757-899X/992/1/011001

Organizing Secretary	**Dr.C.Thiagarajan,** Professor & Head, Department of Automobile Engineering, Saveetha School of Engineering
Joint Secretaries	**Dr.A.Pandiyan**, Associate Professor, Department of Mechanical Engineering, Saveetha School of Engineering
	Mr.A.Muniappan, Associate Professor, Department of Mechanical Engineering, Saveetha School of Engineering
Treasurer	**Mr.D.Vinodh,** Assistant Professor, Department of Mechanical Engineering
	Mr.F.Mohammed Ajmal Sheriff, Assistant Professor, Department of mechanical Engineering

ORGANISING AND PROGRAMME COMMITTEE

Dr.G.Bharathiraja	Dr.K.Shanmuganandam	Dr.N.Karunakaran
Dr.T.Maridurai	Dr.J.Purushothaman	Dr.M.J.Silvister Raju
Dr.S.Madhu	Dr.C.Sadhasivam	Mr.V.Sakthimurugan
Dr.K.Yoganandam	Dr.Deepak Joel Johnson	Dr.D.Banupriya
Dr.B.Navin Kumar	Dr.V.Siva Shankar	Dr.B.Navin Kumar
Dr.S.Krishnaraj	Dr,Melvin Victor Depoures	Dr.K.Archana

IOP Conf. Series: Materials Science and Engineering **992** (2020) 011001 doi:10.1088/1757-899X/992/1/011001

INTERNATIONAL ADVISORY COMMITTEE

Dr.David Jenkins, University of Plymouth, United Kingdom

Dr.Ali M.Abdelsalam, Menoufiya University, Egypt

Dr.Fabio Celani, Sapienza University of Rome, Italy

Dr.Mohamed Thariq Bin Haji Hameed Sultan, Universiti Putra, Malaysia

Dr.Adem KURT, Gazi University, Turkey

Mr.Denis Marchant, Kingston University, United Kingdom

Mr.Abdul Bari.H, Saudi Aramco, Saudi Arabia

Mr.Rabi Mohamed.H, Al Naboodah, Dubai UAE

Dr.Alokesh Pramanik, Curtin University, Australia

Dr.Ancha Venkata Ramayya, Jimma University, Ethiopia

Dr.J.Paulo Davim, University of Aveiro, Portugal

Dr.Christina G.Georgantopoulou, Bahrain Polytechnic, Bahrain

Dr.J.Sadhik Basha, International Maritime College Oman, Sultanate of Oman

IOP Conf. Series: Materials Science and Engineering **992** (2020) 011001 doi:10.1088/1757-899X/992/1/011001

NATIONAL ADVISORY COMMITTEE

IOP Conf. Series: Materials Science and Engineering **992** (2020) 011001 doi:10.1088/1757-899X/992/1/011001

KEYNOTE ADDRESS

Dr.R.VELRAJ,
Professor,
Institute for Energy Studies,
College of Engineering, Guindy,
Anna University, Chennai, Tamil Nadu, India.
E.mail: velrajr@annauniv.edu

Title	Design & Simulation of Thermal Systems
YouTube Link for Keynote address	https://youtu.be/2lyhmtgqZUI

IOP Conf. Series: Materials Science and Engineering **992** (2020) 011001 doi:10.1088/1757-899X/992/1/011001

KEYNOTE ADDRESS

Prof. GYORGY SZEKELY
Advanced Membranes & Porous Materials Center,
King Abdullah University of Science & Technology (KAUST),
Saudi Arabia

E.mail: gyorgy.szekely@kaust.edu.sa

Title	Robust Polymer Materials for Nanofilteration Processess
YouTube Link for Keynote Address	https://youtu.be/nCyM5iCETzQ

IOP Conf. Series: Materials Science and Engineering **992** (2020) 011001 doi:10.1088/1757-899X/992/1/011001

KEYNOTE ADDRESS

Dr .WEIPING WU
George Daniels Lecturer (Assistant Professor)
University of London, London, UK

E.mail: Weiping,Wu@city.ac.uk

Title	Functional Materials for Optoelectronic and Energy Applications
YouTube Link for Keynote Address	https://youtu.be/wARNHjFytbI

IOP Conf. Series: Materials Science and Engineering **992** (2020) 011002 doi:10.1088/1757-899X/992/1/011002

Peer review declaration

All papers published in this volume of *IOP Conference Series: Materials Science and Engineering* have been peer reviewed through processes administered by the Editors. Reviews were conducted by expert referees to the professional and scientific standards expected of a proceedings journal published by IOP Publishing.

- **Type of peer review** : Single-blind
- **Conference submission management system** : EasyChair, icdams20@gmail.com
- **Number of submissions received** : 30
- **Number of submissions sent for review** : 25
- **Number of submissions accepted** : 17
- **Acceptance Rate (Number of Submissions Accepted / Number of Submissions Received X 100)** : 57
- **Average number of reviews per paper** : 2
- **Total number of reviewers involved** : 6
- **Any additional info on review process** :

  All the submission went through single-blind review, based on recommendation from the subject experts (reviewers) and editors the decision communicated to the authors. Based on the recommendation, the authors revised the articles and submitted their revised papers along with response files to the reviewer comments. The revised submission verified by the reviewers and editors for their final recommendation for the submission.

- **Contact person for queries** :

  Dr.T.SHAAFI, Convener, ICDAMS 2020, Professor & Head, Department of Agricultural Engineering, Saveetha School of Engineering, Saveetha Institute of Medical and Technical Sciences, Chennai. Tamilnadu, India. E-mail: icdams20@gmail.com

IOP Conf. Series: Materials Science and Engineering **992** (2020) 012001 doi:10.1088/1757-899X/992/1/012001

Optical analysis of T-shaped stiffened plate under compressive loading using central composite design scheme

Letsatsi M T[1], **Agarwal A**[2]

[1] Department of Industrial Design and Technology, University of Botswana, Gaborone, Botswana

[2] Department of Mechanical Engineering, University of Botswana, Gaborone, Botswana

Email: matlhogonolo.letsatsi@mopipi.ub.bw

Abstract. The buckling and post-buckling strength of stiffened panels subjected to compressive loads largely depends upon type and dimensions of stiffeners used. The current research investigates the use of T shaped stiffener in vertical configuration in rectangular plate with square opening for improving strength and buckling characteristics. The static structural analysis is conducted using ANSYS FEA software and dimensions of T shaped stiffeners are optimized using Taguchi response surface optimization. The design points are generated using CCD (Central Composite Design) scheme of response surface method and corresponding responses of optimization variables are captured in 3D plots. The sensitivity chart is generated to determine the effect of optimization variables on parameters of interest which are stress, strain, deformation, buckling load and safety factor. The optimization technique has successfully provided set of values for different parameters for which buckling and stresses are minimum and maximum.

1. Introduction

The stiffened panels possess high structural efficiency and therefore it is used in aircraft fuselage as it demands higher strength and demands higher load bearing capability. In case of heavy loading where web enter post-buckling regime the stiffener and skin tend to separate. In plate like buckling, longitudinal edges of the stiffened steel plates are considered as simply supported. The advantage of such design is development of local buckling prior to global/web buckling of entire panel and the optimized dimensions of stiffeners could significantly increase the buckling resistance of stiffened plate. N. Raghu Prasad, Jeeoot Singh [1] conducted FEA analysis on stiffened plates with and without openings. The boundary condition involved simply supported ends with edge compression. The numerical results shown minor variations in magnitude and pattern of stress with opening as compared to plate without opening. Ali Reza Pouladkhan, Jalil Emadi, Majid Safemehr [2] conducted FEA analysis on thin plate using Abacus software. The buckling loads were evaluated from equilibrium equation with mesh convergence studies. Figure 1 shows application of longitudinal and transverse stiffener in rectangular plate.

IOP Conf. Series: Materials Science and Engineering **992** (2020) 012001 doi:10.1088/1757-899X/992/1/012001

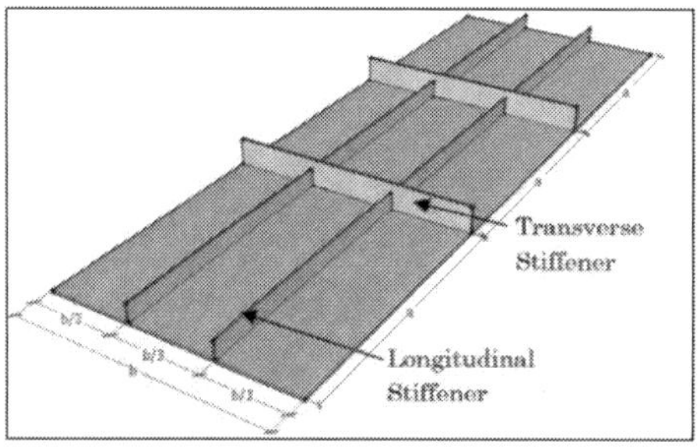

Figure 1. Longitudinal and transverse stiffener in rectangular plate

The analytical results were in close agreement with FEA results. Sang-Rai Cho, Hyun-seung Lee [3] conducted experimental study of 33 stiffened plate subjected to lateral collision. The buckling pattern was evaluated and the kinetic energy and potential energy obtained for experimental results were validated with analytical results. Ghania Ikhenazen, Messaoud Saidani [4] investigated buckling behavior of isotropic plate subjected to patch loads using total energy concept. The results obtained are validated with FEA results and graph was generated for buckling factor vs aspect ratio.

Byklum et al. [5] devised a computational model for analysing local buckling of stiffened panels and is more accurate than finite element method which takes in to account any combination of biaxial loads (both tensile and compressive).Houlston et al. [6][7][8] investigated the application of finite strip method on stiffened panels and validation is made using results obtained from ADINA software and experimental data. Nurick et al. [9] and Rudrapatna et al. [10] conducted testing on fully built-in stiffened square plates subjected to blast loading. The experimental analysis has shown tearing of plates at fixed boundary for smaller stiffener. Louca and Pan [11][12] have analysed stiffened plate using DYNA 3D finite element software to determine local and global response of stiffened plate when subjected to blast loading. The results have shown that structural response of stiffened plate is not influenced by direction of blast loading and type of stiffeners.

2. Optimize design of dimensions of T shaped stiffeners

The objective of this research is to optimize design of dimensions of T shaped stiffeners used in rectangular plate subjected to vertical loading conditions. The material used for shear plates is structural steel and optimization parameters are beam lower with and beam upper width of T shaped stiffeners. The optimization is done using Taguchi response surface method and structural analysis is conducted using Finite Element Method [13].

The structural analysis of rectangular plate is done using finite element method (FEM) which is based on discretization of domain. The analysis has three stages i.e. pre-processing, solution and post-processing. The first stage is CAD modelling using ANSYS software. ANSYS design modeller is specific tool used for designing and editing operation. The CAD model of rectangular plate is modelled in ANSYS design modeller from schematic as shown in figure 2.

IOP Conf. Series: Materials Science and Engineering **992** (2020) 012001 doi:10.1088/1757-899X/992/1/012001

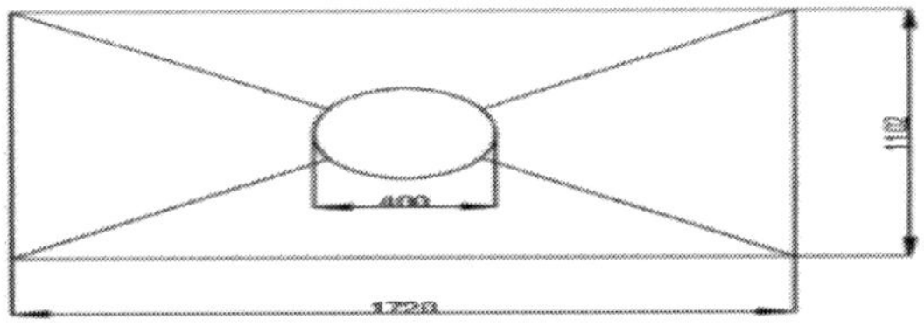

Figure 2. Schematic of rectangular plate with hole

The 2-dimensional CAD model is developed and the opening is changed from circular to square and stiffeners modelled is T shaped as shown in figure 3.

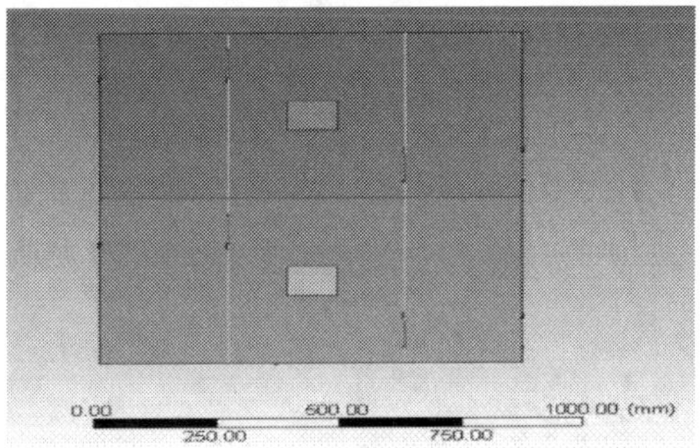

Figure 3. CAD modelling of rectangular plate with square opening

The model is meshed using tetra elements of appropriate size and shape as shown in figure 4 below. After meshing, appropriate loads and boundary conditions are assigned. The bottom face is fixed and top face is applied with 900kN vertical load as shown in figure 5.

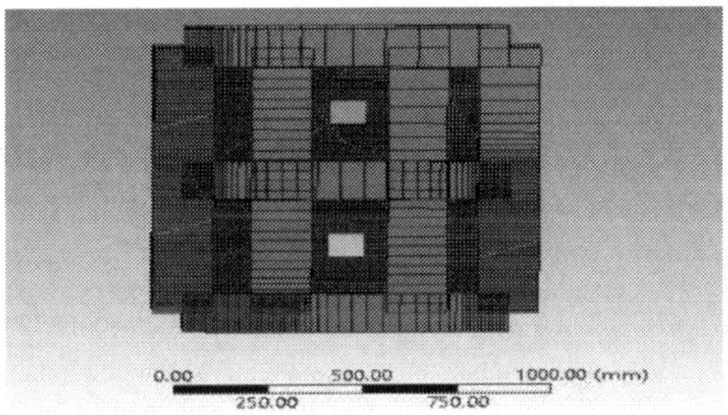

Figure 4. Meshed model of rectangular plate with square opening

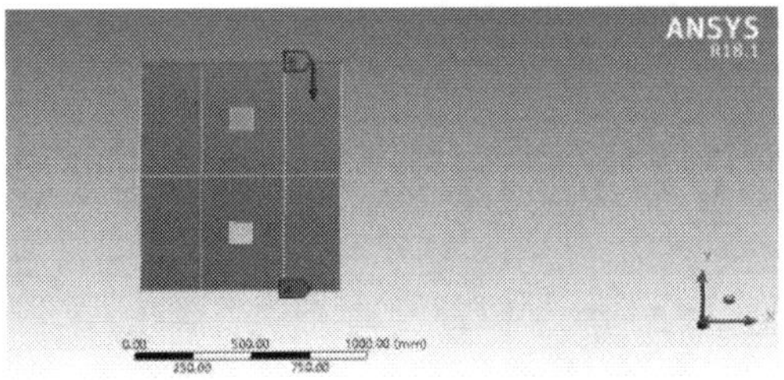

Figure 5. Loads and Boundary conditions

The simulation is run with sparse matrix solver and values of output variables are determined in solution stage.

IOP Conf. Series: Materials Science and Engineering **992** (2020) 012001 doi:10.1088/1757-899X/992/1/012001

3. Results and discussion

The contour plots of output parameters (deformation, principal stress, principal strain and safety factor) generated from FEA analysis to get distribution pattern. These contour plots showed critical regions of rectangular plate. The top most beam of rectangular plate structure has maximum deformation as shown in figure 6 and top left end of bottom plate has maximum principal stress as shown in figure 7.

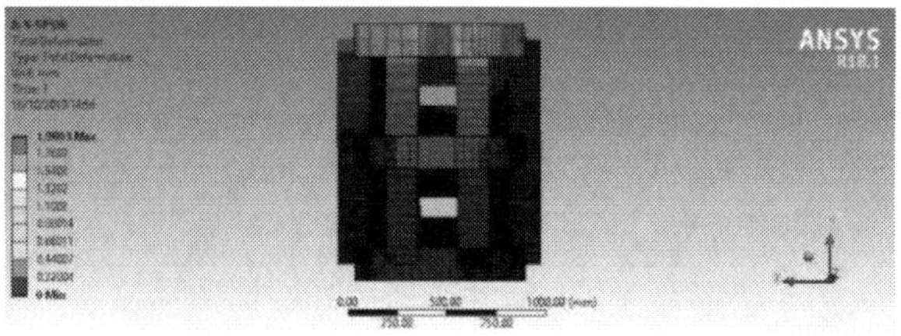

Figure 6. Deformation plot for square opening with T beams

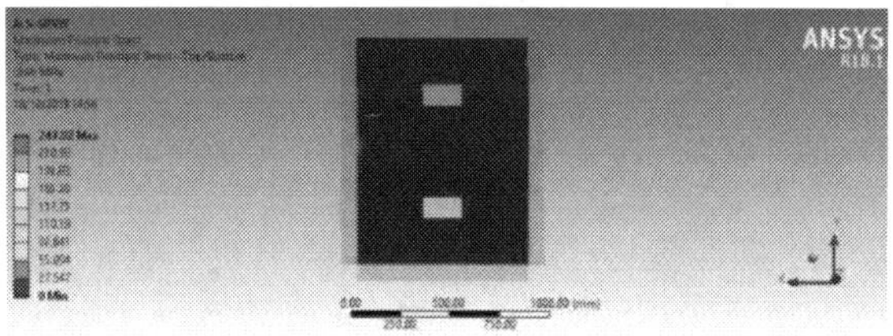

Figure 7. Max Principal stress for square opening with T beams

The safety factor is high for vertical members, horizontal members lowest is observed for bottom left and bottom right portion of plate as shown in figure 8.

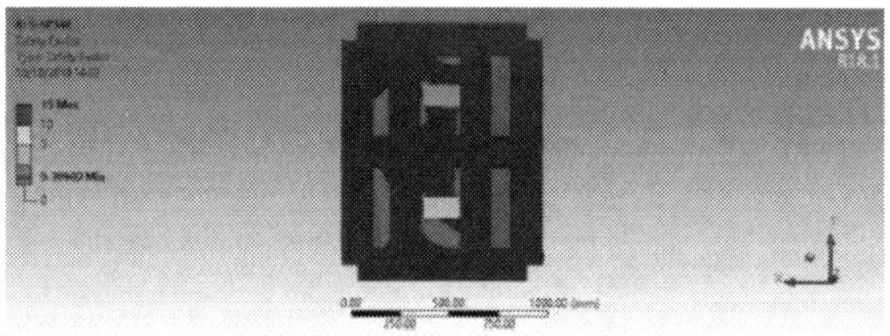

Figure 8. Safety factor for square opening with T beams

The deformation load multiplier plot is generated to determine critical buckling load of structure as shown in figure 9. The plot shows the region immediately above bottom plate opening are susceptible to buckling (shown by red colour).

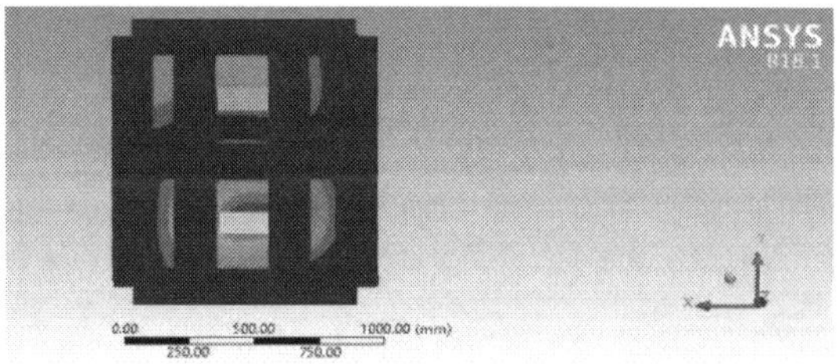

Figure 9. Total deformation load multiplier for square opening with T beams

The responses which are output variable are optimized on the basis of input variable which are independent. The independent variables are denoted by $x_1, x_2, x_3.....x_n$. These variables are independent and controlled by experimenter with negligible error. The relationship between independent variable and dependent variable can be expressed as-

$$Y = f(x_1,x_2,x_3.....x_n) + \epsilon \tag{1}$$

Where ϵ is error observed in the response y.The two variable regression model of quadratic response is given by

$$y=\beta_0 + \beta_1x_1 + \beta_2x_2 + \beta_3x_1^2 + \beta_4x_2^2 + \beta_5x_1x_2 \tag{2}$$

Where β_0, β_1, β_2, β_3, β_4 and β_5 are the regression coefficients. The response surface optimization of the current research deals with 2 design parameters which are side width (x_1) and lower width (x_2). To establish the cause and effect relationship design of experiments (DOE) is performed using ANSYS software. These experiments are conducted using central composite design method or CCD. The design points are generated from DOE is shown in table 1 below.

Table 1. Design Points generated using Taguchi Design of Experiments

A Name	B P17-sidewidth (mm)	C P18-lowerwidth (mm)	D P1-Total Deformation Load Multiplier	E P5-Total Bending moment max (N mm)	F P10-Total Deformation max (mm)	G P11-Max Principal Stress (MPa)	H P12-Max Principal Elastic Strain max mm mm^{-1}	I P13-Safety Factor Min
1	159	161	70.479	1.0223E+o7	1.9358	247.99	0.0012282	0.30915
2	158	161	7.0241	1.0227E+o7	1.9128	248.65	0.0012316	0.30811
3	160	161	7.0714	1.0219E+o7	1.9592	247.33	0.00U249	0.31019
4	159	160	7.0436	1.0197E+o7	1.9567	248.63	0.0012309	0.30798
5	159	162	7.0522	1.0249E+o7	1.9153	247.36	0.00 U255	0.31033
6	158	160	7.0198	1.0201E+o7	1.9333	249.29	0.0012343	0.30693
7 DP 0	160	160	7.0702	1.0!93E+o7	1.9803	247.92	0.0012273	0.30909
8	158	162	7.0285	1.0253E+o7	1.8925	248.01	0.00 U289	0.30929
9	160	162	7.0757	1.024SE+o7	1.9384	246.7	0.0012222	0.31136

The graph of deformation vs side width as shown in figure 10 depicts an increase in total deformation load multiplier or buckling load with increase in side width of T beams up-to maximum value of 160mm.

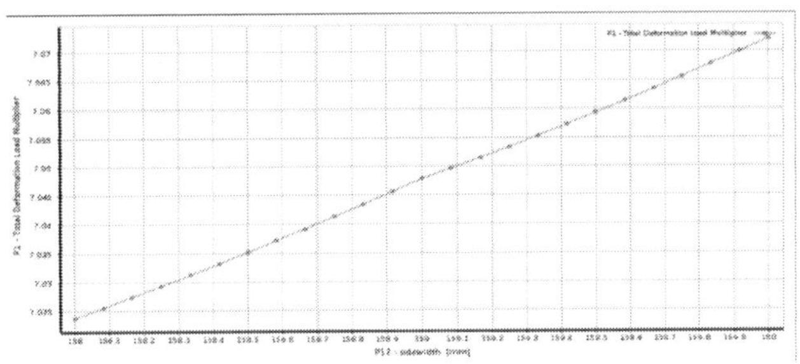

Figure 10. Deformation vs side width of square opening with T beams

The graph of deformation vs lower width as shown in figure 11 depicts increase in total deformation load multiplier or buckling load with increase in lower width of T beams up-to maximum value of 16mm.

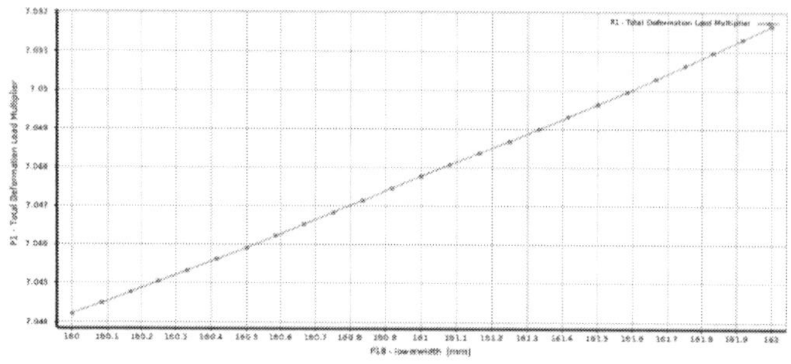

Figure 11. Deformation vs lower width of square opening with T beams

The 3D response surface plot of load deformation multiplier is shown in figure 12. The deformation load multiplier has highest value for side width range from 160mm to 162mm and lower width ranging from 160mm to 162mm and has lowest value for side width range from 158mm to 158.5mm and lower width ranging from 161.5mm to 160mm.

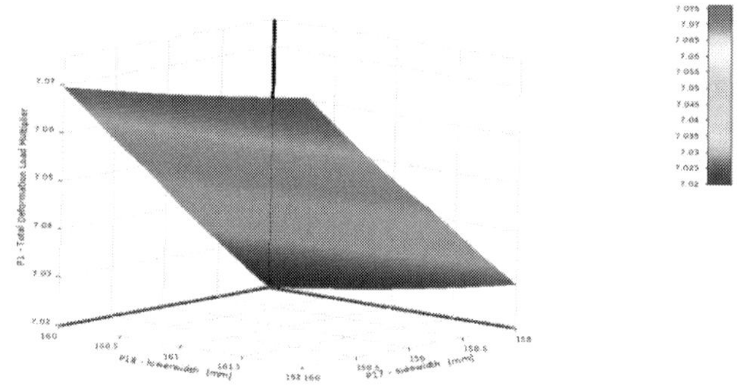

Figure 12. 3D Response Surface Plot of square opening with T beams

The graph of deformation vs side width as shown in figure 13 depicts the decrease in total bending moment with increase in side width of T beams up-to maximum value of 160mm.

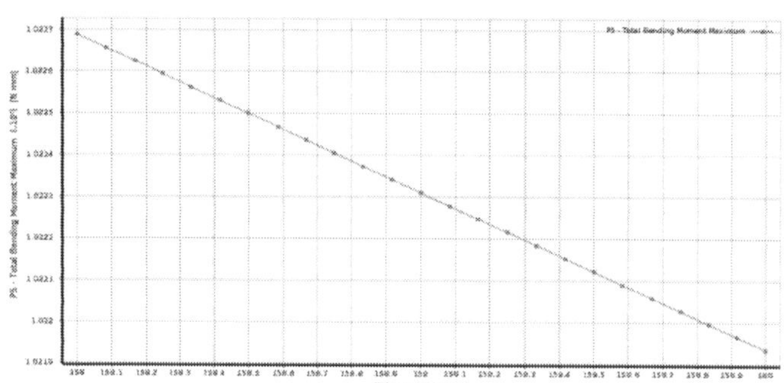

Figure 13. Bending moment vs side width of square opening with T beams

The graph of deformation vs side width as shown in figure 14 depicts the increase in total bending moment with increase in lower width of T beams upto maximum value of 162mm.

IOP Conf. Series: Materials Science and Engineering **992** (2020) 012001 doi:10.1088/1757-899X/992/1/012001

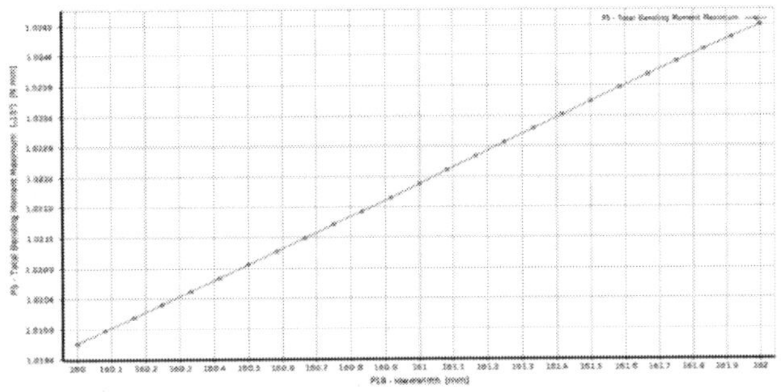

Figure 14. Bending moment vs lower width of square opening with T beams

The 3D response surface plot of bending moment is shown in figure 15. The total bending moment has highest value for side width range from 158mm to 162mm and lower width ranging from 161.5mm to 162mm and has lowest value for side width range from 158mm to 160mm and lower width ranging from 160mm to 160.5mm.

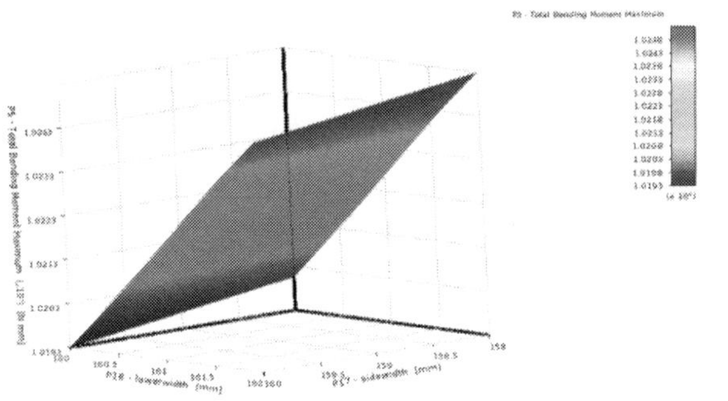

Figure 15. 3D Response Surface Plot of square opening with T beams

The graph of deformation vs side width as shown in figure 16 depicts increase in total deformation with increase in side width of T beams up-to maximum value of 160mm. The graph of deformation vs lower width is shown in figure 17 depicts decrease in total deformation with increase in lower width of T beams up-to maximum value of 162mm.

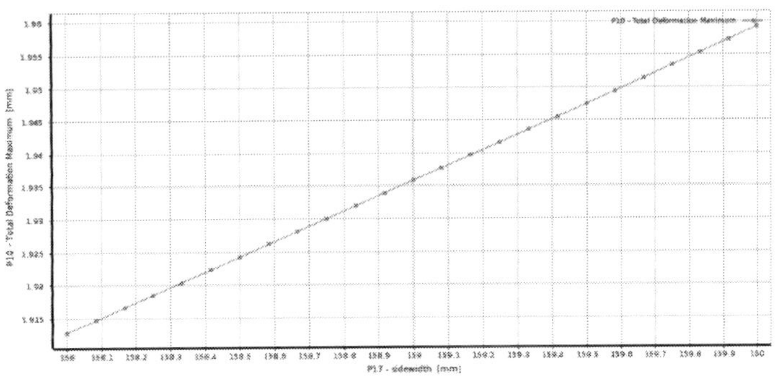

Figure 16. Total deformation vs side width of square opening with T beams

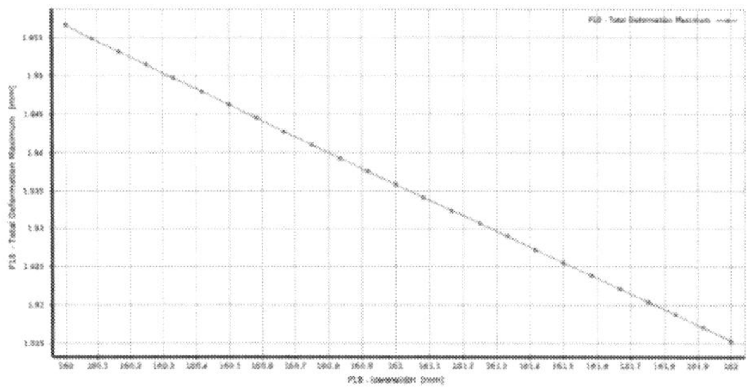

Figure 17. Total deformation vs lower width of square opening with T beams

The 3D response surface plot of total deformation is shown in figure 18, which shows highest value of deformation for side width range from 159.5mm to 160.62mm and lower width ranging from 160mm to 160.5mm and has lowest value of deformation for side width range from 158mm to 158.5mm and lower width ranging from 161.5mm to 162mm.

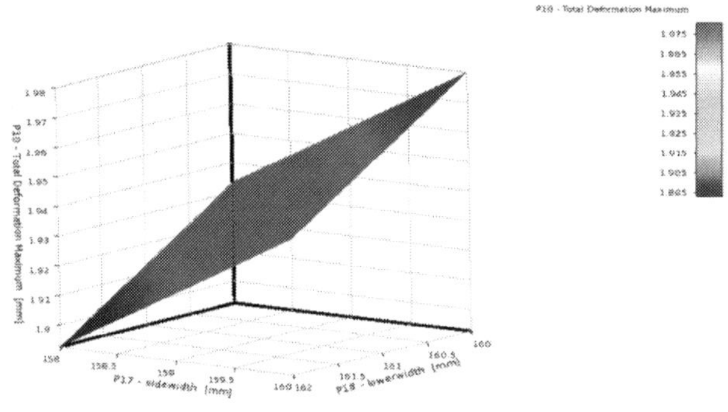

Figure 18. 3D Response Surface Plot of square opening with T beams

The graph of Maximum principal stress vs side width as shown in figure 19 depicts the decrease in Maximum principal stress with increase in side width of T beams up-to maximum value of 160mm. The variation of Maximum principal stress vs lower width shows the decrease in maximum principal stress with increase in lower width of T beams up-to maximum value of 162mm.

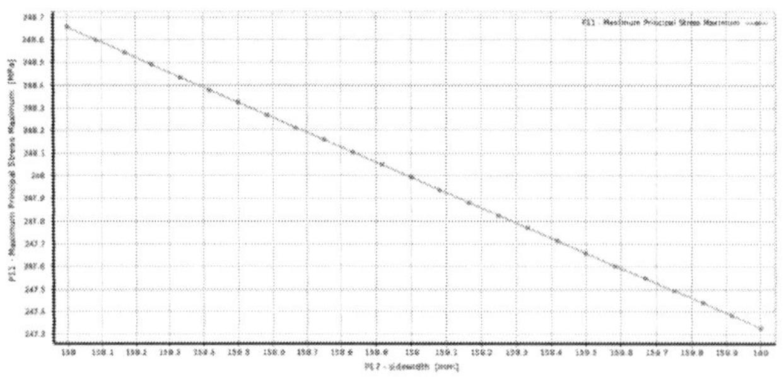

Figure 19. Maximum principal stress vs side width of square opening with T beams

The 3D Response Surface Plot of principal stress is shown in figure 20. The plot shows the maximum principal stress has highest value for side width range from 158mm to 158.5mm and lower width ranging

IOP Conf. Series: Materials Science and Engineering **992** (2020) 012001 doi:10.1088/1757-899X/992/1/012001

from 160mm to 160.5mm and has lowest value for side width range from 159.5mm to 160mm and lower width ranging from 161.5mm to 162mm.

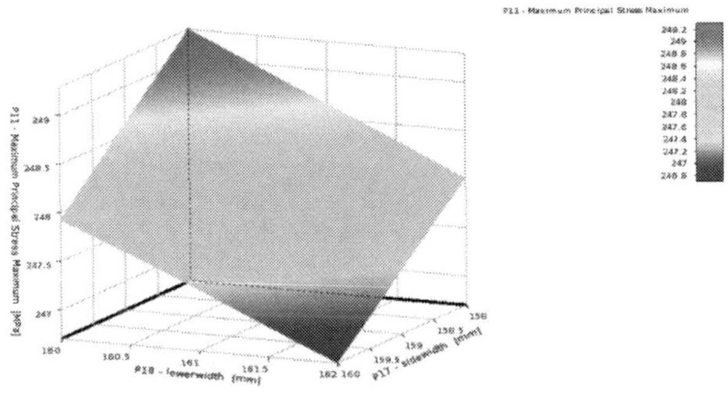

Figure 20. 3D Response Surface Plot of square opening with T beams

The graph of Maximum principal stress vs side width shows decrease in Maximum principal strain with increase in side width of T beams up-to maximum value of 160mm.

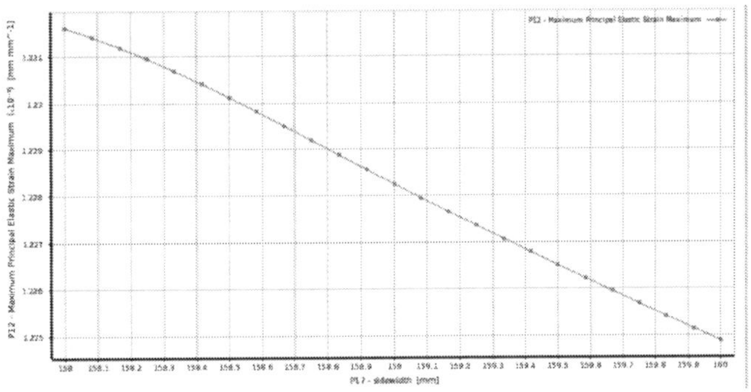

Figure 21. Maximum principal stress vs side width of rectangular opening with T beams

The graph of Maximum principal stress vs lower width as shown in figure 22 depicts decrease in maximum principal stress with increase in lower width of T beams up-to maximum value of 162mm.

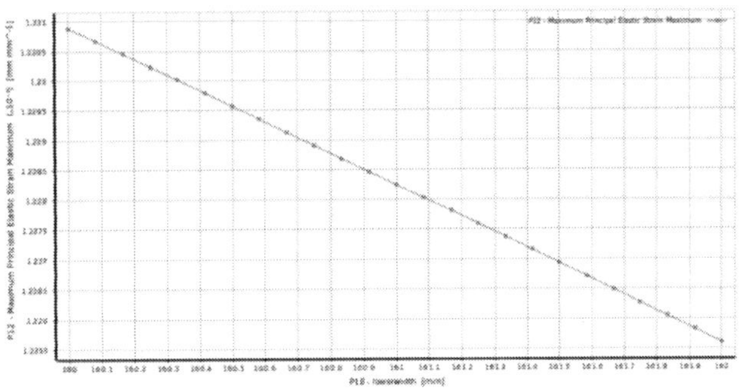

Figure 22. Maximum principal strain vs lower width of square opening with T beams

The 3D Response Surface Plot of principal elastic strain with T beams is shown in figure 23. The plot shows maximum principal strain has highest value for side width range from 158mm to 158.5mm and

IOP Conf. Series: Materials Science and Engineering **992** (2020) 012001 doi:10.1088/1757-899X/992/1/012001

lower width ranging from 160mm to 160.5mm and has lowest value for side width range from 159.5mm to 160mm and lower width ranging from 161.5mm to 162mm.

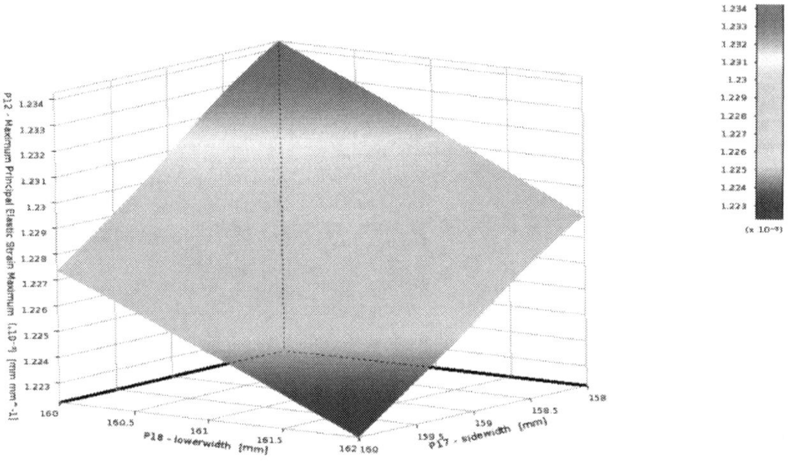

Figure 23. 3D Response Surface Plot of square opening with T beams

The graph of safety factor vs side width as shown in figure 24 depicts increase in safety factor with increase in side width of T beams up-to maximum value of 160mm. The variation of safety factor vs lower width depicts increase in safety factor with increase in lower width of T beams up-to maximum value of 162mm.

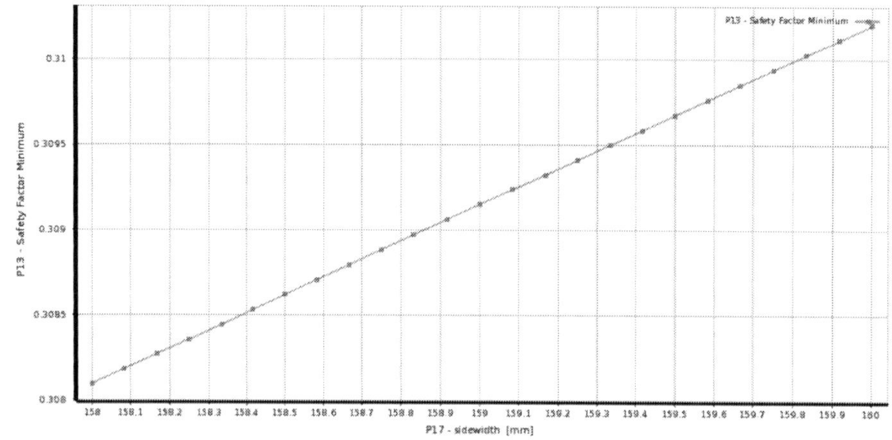

Figure 24. Safety factor vs side width of square opening with T beams

The safety factor plot generated is shown in figure 25. The plot shows that the highest value of safety factor for side width range from 159.5mm to 160mm and lower width ranging from 161.5mm to 162mm and lowest value of safety factor for side width range from 158mm to 158.5mm and lower width ranging from 160mm to 160.5mm.

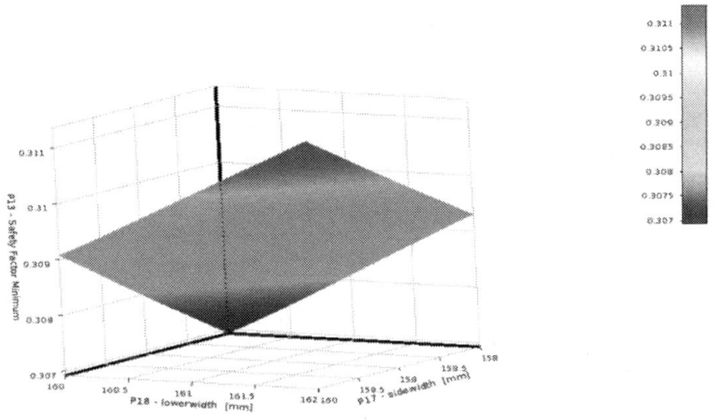

Figure 25. 3D Response Surface Plot of square opening with T beams

The sensitivities of different optimization variables are studied to determine its effect on output variables as shown in figure 26. This would enable designer/engineer to identify the optimization variable which could cause maximum variation in output variable (the one having higher sensitivity percentage).

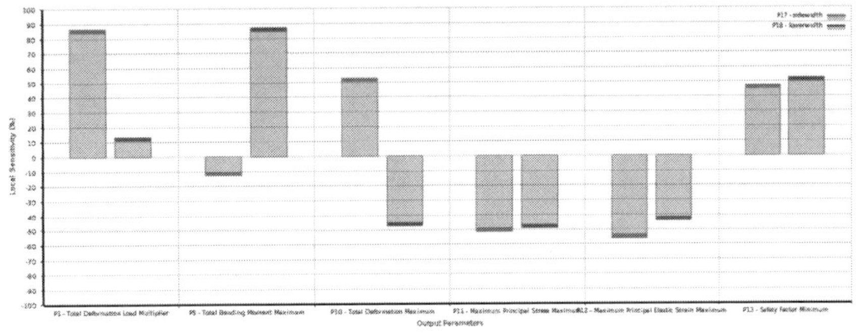

Figure 26. Sensitivity plot of various parameters for square opening with T beams

4. Conclusion

FEA analysis conducted on stiffened rectangular plates with square opening. The stress distribution and buckling characteristics of rectangular plate has shown critical areas of failure. The dimensions of I shaped stiffeners are optimized and specific set of optimization variable values are generated. The sensitivity plots of variables are also generated which shows the effect of different optimization variables on output variables. The detailed conclusion points are as follows:

1. For total deformation load multiplier variable both beam lower width and beam side width shows positive sensitivities. Beam lower width sensitivity percentage is 13.32 (positive) and beam side width sensitivity percentage is 86.52 (positive). Therefore, beam side width has higher effect on deformation load multiplier or buckling load.
2. For bending moment beam side width sensitivity percentage is 12.61 (negative) and beam lower width sensitivity percentage is 87.38 (positive). Therefore, beam lower width has higher effect on sensitivity of bending moment.

3. For total deformation beam lower width sensitivity percentage is 47.09 (negative) and beam side width sensitivity percentage is 52.89(positive). Therefore, beam side width has higher effect on total deformation of beam.
4. For maximum principal elastic strain beam side width sensitivity percentage is 51.15(negative) and beam lower width sensitivity percentage is also 48.76 (negative). Therefore, beam side width has higher effect on maximum principal elastic strain.
5. For safety factor beam side width sensitivity percentage is 47.44 (positive) and beam lower width sensitivity percentage is also 52.53 (positive). Therefore, beam lower width has more effect on safety factor.
6. The optimum results are obtained for beam side width ranging from 159.5 to 160mm and beam lower width ranging from 161.5mm to 162mm.

References

[1] Prasad N and Singh J 2015 Buckling Analysis Of Rectangular Plates With Cutouts And Partial Edge Compression *IARJSET* 2 126–9
[2] Pouladkhan A R 2011 Numerical Study of Buckling of Thin Plates
[3] Cho S-R and Lee H-S 2009 Experimental and analytical investigations on the response of stiffened plates subjected to lateral collisions *Mar. Struct.* **22** 84–95
[4] Ikhenazen G, Saïdani M and Chelghoum A 2010 Finite element analysis of linear plates buckling under in-plane patch loading *J. Constr. Steel Res.* **66** 1112–7
[5] Byklum E and Amdahl J 2002 A Simplified Method for Elastic Large Deflection Analyses of Plates and Stiffened Panels due to Local Buckling *Thin-Walled Struct.* **40** 925–53
[6] Houlston R and DesRochers C G 1987 Nonlinear structural response of ship panels subjected to air blast loading *Comput. Struct.* **26** 1–15
[7] Houlston R 1989 Finite strip analysis of plates and stiffened panels subjected to air-blast loads *Comput. Struct.* **32** 647–59
[8] Houlston R and Slater J E 1987 A summary of experimental results on square plates and stiffened panels subjected to air-blast loading *Shock and Vibration Information Center The Shock and Vibration Bulletin. Part 2: Instrumentation, Shock Analysis and Shock Testing p 55-67(SEE N 88-17053 09-39)*
[9] Nurick G N, Olson M D, Fagnan J R and Levin A 1995 Deformation and tearing of blast-loaded stiffened square plates *Int. J. Impact Eng.* **16** 273–91
[10] Rudrapatna N S, Vaziri R and Olson M D 2000 Deformation and failure of blast-loaded stiffened plates *Int. J. Impact Eng.* **24** 457–74
[11] Louca L A, Pan Y G and Harding J E 1998 Response of stiffened and unstiffened plates subjected to blast loading *Eng. Struct.* **20** 1079–86
[12] Pan Y and Louca L A 1999 Experimental and numerical studies on the response of stiffened plates subjected to gas explosions *J. Constr. Steel Res.* **52** 171–93
[13] Agarwal A, Seretse O M and Pumwa J 2020 Finite element and Taguchi response analysis of the application of graphite aluminum mmc in automotive leaf spring *J. Mech. Contin. Math. Sci.* **15** 168–79

Feasibility study of using permanent magnets as indirect temperature monitor in irradiation capsule of nuclear research reactor

Suman Saurav [*1], M. Muthuganesh[2], P. K. Chaurasia[2] and S. Murugan[2]

[1]Homi Bhabha National Institute, Indira Gandhi Centre for Atomic Research
[2]Reactor Design and Technology Group, Indira Gandhi Centre for Atomic Research, Kalpakkam - 603102, India

[*]Email: sumansaurav@igcar.gov.in

Abstract. In sodium cooled nuclear fast reactors, materials used for fuel cladding and structural components can undergo significant physical and dimensional changes due to exposure to high energy neutrons. At high temperature and radiation conditions prevailing in the nuclear reactor, material properties change considerably. Information about the temperature of exposure during specimen irradiation in a nuclear reactor is important since the temperature is having a strong influence on the mechanical and physical properties of materials. A proper evaluation of the effects due to radiation on materials needs a precise knowledge of the irradiation temperature. Many times, standard techniques for measuring temperature are not practicable in reactors due to the availability of limited instrumented positions. Therefore, the use of small-sized indirect temperature monitors such as melt wires, permanent magnets, etc. which can be incorporated along with the samples and examined after the irradiation are preferred. In the present study, permanent magnets of neodymium grade N38 and N38SHX of different sizes have been studied for their possible use as an indirect temperature monitor. The study was carried out first for their compatibility with sodium since the monitor will be exposed to sodium during irradiation experiments in fast reactors, and then by exposing to different temperatures in an electrical furnace for the determination of Curie point to study the feasibility of using these materials as indirect temperature monitors. Since the material loses its magnetic properties when the temperature of exposure exceeds the Curie point, by using a series of magnets having progressively increasing curie points, the temperature of exposure can be determined by testing these magnets for the retention of their magnetic properties. Specifically, since there is variation in the Curie temperature as reported in the literature, the experimental method was developed to determine the Curie temperature accurately. This paper discusses the details of the experiments carried out in our laboratory and the results obtained from these studies.

Keywords: Neodymium magnet, Irradiation device, Temperature detector, Demagnetization, Curie point

1. Introduction

Fast Breeder Test Reactor (FBTR) at Indira Gandhi Centre for Atomic Research (IGCAR), Kalpakkam, India offers good facilities for irradiation of structural material specimens to determine the change in the properties due to irradiation [1, 2]. In the reactor, the irradiation performance of the materials will be influenced by the neutron flux level and temperature [3, 4]. It is essential that the temperature of specimens during irradiation is determined accurately for correct interpretation of results obtained in the irradiation experiments. However, many times the standard techniques for measuring temperature such as the use of thermocouple [5] are not practicable in the experimental facilities of nuclear reactor core, due to the non-availability of instrumented irradiation locations. The general practice is the use of small size indirect temperature indicators in the irradiation capsule along with the sample and to examine it after the irradiation experiment. Some indirect techniques available for measuring temperature include the use of melt wire [6], Silicon Carbide (SiC) temperature monitor [7], and thermal expansion difference (TED) monitor [8]. Melt wire method is used to measure the peak temperature attained using wires of known composition and melting temperature. The multiple wires are encapsulated into a single compact unit. This method requires the identification of the melting of wire unambiguously after the irradiation experiment. TED method is used to measure the peak temperature by a change in the dimension of the tube (fully filled with liquid such as sodium) when exposed to high temperatures. In the TED method, there is a chance for improper filling of sodium, so it can give an inaccurate measurement of the temperature.

In the present work, a magnetic method (Curie point technique) for peak temperature monitoring has been explored. In this technique, the temperature can be measured using magnetic materials. In theory, if the temperature of the magnet exceeds its Curie point (T_C), the magnet loses its magnetic property [9]. The phenomenon of the loss of magnetic property concerning Curie temperature has been studied using an in-house experimental setup to explore this phenomenon and to identify the peak temperature attained. It is based on the principle that when the temperature of the magnet exceeds the Curie temperature the magnet loses the magnetic properties. Conversely, if it is not so, it can be inferred that the maximum temperature remained below the Curie point. It is reported in the literature [10] that a similar work has been carried out by Marth et al. [11] in Karlsruhe Germany but not much detail is available in the open literature. Using a succession of the magnets with different Curie points it is possible to attain sufficient data to determine the maximum temperature of specimens at various partitions of an irradiation capsule in the reactor during irradiation.

2. Selection of magnetic materials for temperature monitor

For the measurement of the peak temperature in the irradiation capsule the magnets can be selected based on following properties: (a) They should have Curie point near to the temperature of interest, (b) They should have good resistance to radiation and, (c) It is preferred that they are compatible with liquid sodium, the coolant of the fast reactor. There are various types of permanent magnets available having different Curie points. By varying the composition of the magnetic material, different Curie point temperatures can be achieved [12-14]. Radiation effect on the permanent magnet is well studied and permanent magnets, such as neodymium magnet, Sm-Co, Alnico, have been reported for higher tolerance to radiation [15-17]. As part of the feasibility study as the temperature monitor in this work, neodymium magnets of grade N38 and grade N38SHX have been selected in different sizes and shapes (Figure 1) for the experiment. The neodymium magnet has a Curie point near to our temperature of interest for the irradiation experiment, and these magnets are also easily available. Compositions of these magnets are shown in Table 1.

Table 1: Composition of magnets used in the study

Magnet	Composition (percent by weight)					
	Neodymium	Iron	Boron	Aluminium	Niobium	Dysprosium
Neodymium magnets	29-32	64-68.5	1.0-1.2	0.2 - 0.4	0.5- 1	0.8- 1.2

IOP Conf. Series: Materials Science and Engineering **992** (2020) 012002 doi:10.1088/1757-899X/992/1/012002

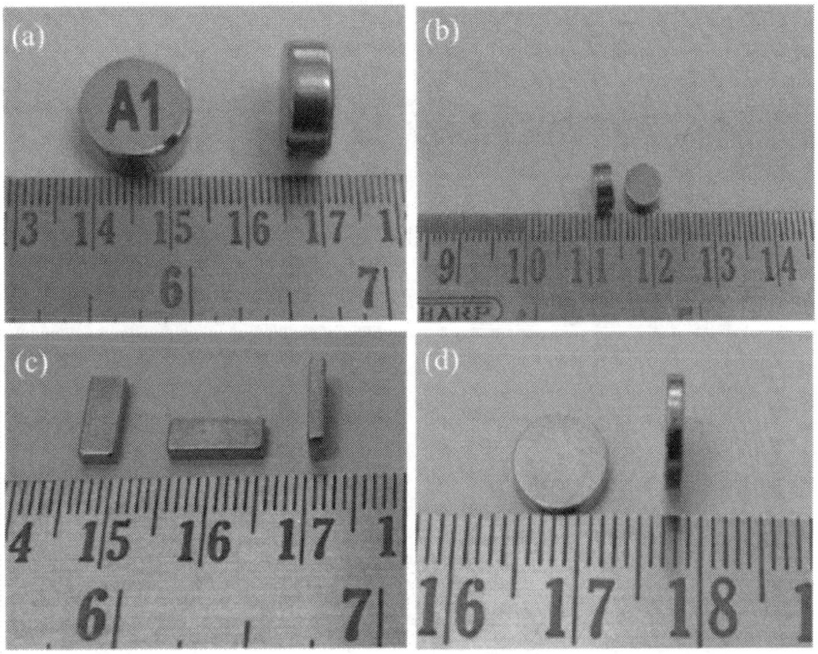

Figure 1: Grade N38 Neodymium magnet (a) 15mm diameter X 5mm width (b) 5mm diameter X 3mm width; Grade N38SHX Neodymium magnet (c) 10mm X 4mm X 2mm (d) 8mm diameter X 1mm width

3. Sodium compatibility test of magnet samples

For the measurement of exposure temperature of the specimens in the non-instrumented irradiation capsules, the magnet samples are being studied for their use as the indirect temperature monitor.The irradiation capsule, where the specimens are kept, is normally filled with the sodium for providing the near-uniform temperature distribution within the capsule. Hence, the magnet which is to be used as the temperature monitor should be compatible with the sodium environment.Therefore, it is necessary to check the compatibility of the neodymium magnet with sodium. The compatibility test experiment of the magnet sample in sodium has been carried out in the glove box with an argon environment. The magnet samples were kept in the sodium container as shown in Figure 2 for the duration of about 150 hours at the temperature of 150 °C.

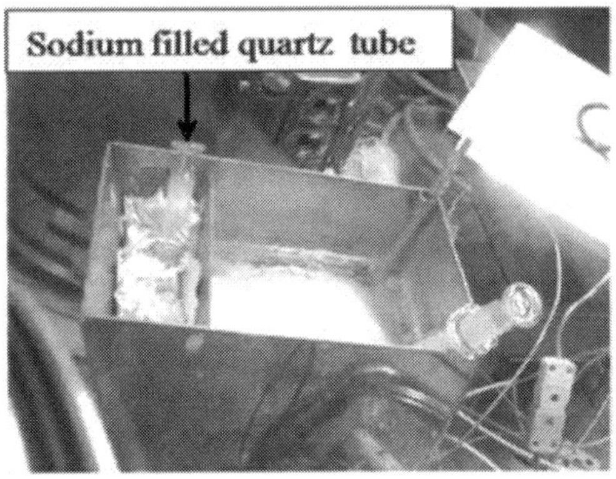

Figure 2: Magnet samples are kept in quartz tube which in turn is placed in sodium container

The magnet samples before and after exposure to the sodium are shown in Figure 3(a) and 3(b) respectively. The weight and the dimensions were measured for the magnet samples before and after exposure to the sodium, and are shown in Table 2.

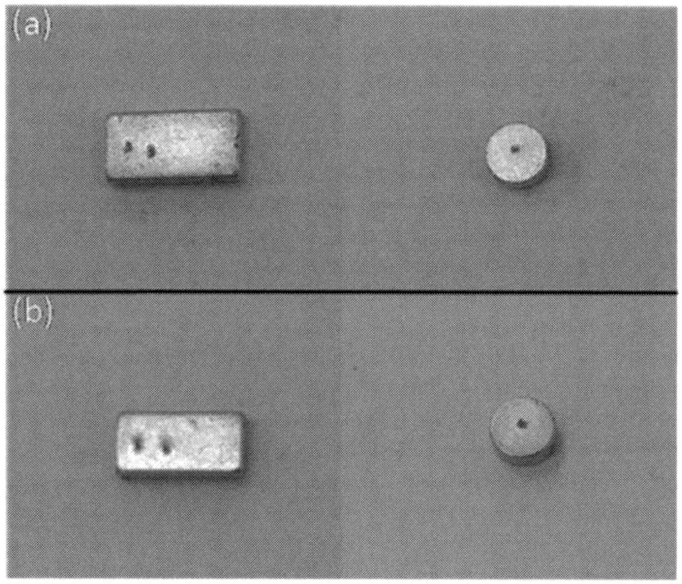

Figure 3: Magnet samples (grade N38, disk type, and grade N38SHX, plate type) (a) before sodium exposure, and (b) after sodium exposure

Table 2: Weight and dimension of magnet samples

Sl. No.	Type of magnet	Weight of magnet (g)		Dimensions	
		before exposure to sodium	after exposure to sodium	before exposure to sodium	after exposure to sodium
1.	Grade N38 (Disk type)	0.556	0.556	5 mm diameter x 3 mm width	5 mm diameter x 3 mm width
2.	Grade N38SHX (Plate type)	0.530	0.530	10 mm x 4 mm x 2 mm	10 mm x 4 mm x 2 mm

The sodium compatibility test results show that there is no significant effect on the magnet due to exposure to sodium and they can be used as temperature monitors in the irradiation capsule in the sodium environment.

4. Determination of Curie point of the magnet samples

Compounds of neodymium magnets contain iron and it is reported that the variation in the iron weight percentage influences Curie temperature. The neodymium magnets (Grade N38 and Grade N38SHX) used in this study contains 64% to 68.5% of iron (by weight) as shown in table 1 and due to this, Curie temperatures of the samples are not known precisely. A similar observation is reported in the literature [18] on the influence of the material composition of cobalt ferrite on curie points while working for stress sensor applications. Hence it is important that before using these magnets as a temperature monitor, an experiment is to be carried out for estimating the Curie point accurately. In this experiment, the Curie temperature of the permanent magnets which can be used as indirect temperature monitors has been determined. An experimental setup (Figure 4) was designed to determine the Curie point of a permanent magnet sample by heating it above the Curie point. A magnetic probe is made by using two ceramic tubes, which are of 0.6m long and two magnetic bases which have higher Curie point is attached at the bottom of the ceramic tubes as shown in Figure 4. These magnetic bases are connected with a battery using conducting wire. The magnetic bases of the probe are also attached with two numbers of K-type thermocouples. The sample magnet material is placed on the same plane in contact with the two magnetic bases, and as a result of which a closed electrical circuit is made. The whole setup is placed in the electric furnace to determine the temperature at which the sample magnet become demagnetized. The data (temperature and voltage) were recorded using a data logger (Yokogawa make) at the rate of 2 data per second. The furnace was started at 30 °C with a step increase in temperature of 5 °C per minute. In this arrangement, the sample magnet functions as a switch in the electrical circuit. When the sample magnet becomes demagnetized due to furnace temperature crossing the Curie point of sample magnet, the circuit would break, and the output voltage would become zero.

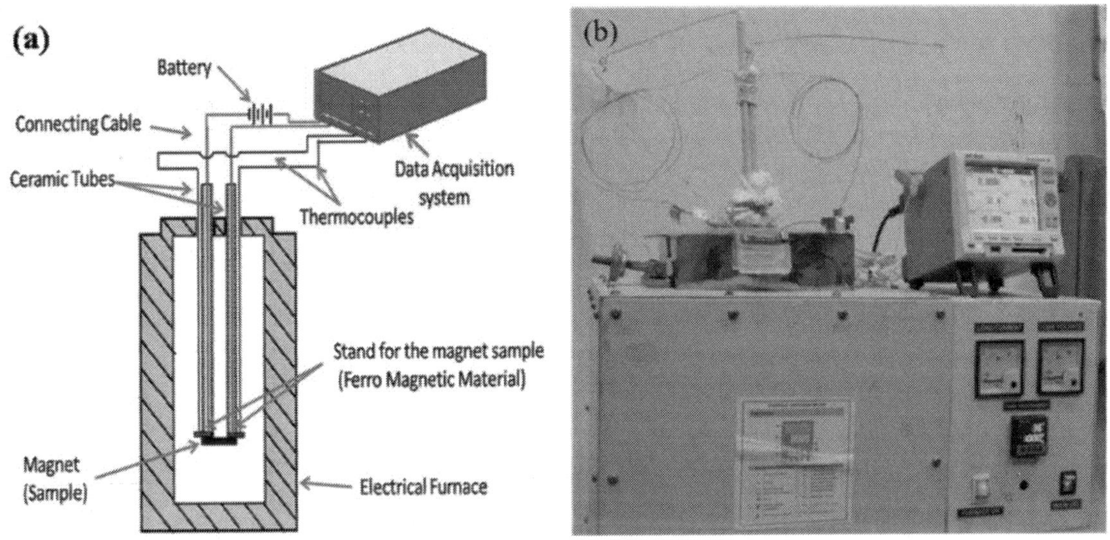

Figure 4: (a) Schematic view of the experimental setup and (b) The actual setup of the experiment

The experimental results for estimating the Curie points of neodymium magnets are shown in Figure 5. It can be seen from Figure 5 that as the temperature increases the magnets lose the magnetic properties and at a particular temperature complete loss in permanent magnetism results in drop down of magnets from the electrical circuit. Eventually, this leads to zero voltage. The Curie point for the tested grade N38 magnets are shown in Figure 5 (a) and (b). It was found to be 255 ± 5 °C (n=10, where n being independent trials) for both the sizes i.e.,15mm diameter x 5mm width (Figure 1(a)) and 5mm diameter x 3mm width (Figure 1(b)). The experimental results of neodymium magnet N38SHX are shown in Figure 5(c) and (d). The N38SHX grade magnet lost its permanent magnetic property at the temperature of 270 ± 5 °C (n=10), for the sizes of 10mm x 4mm x 2mm (Figure 1(c)) and 8mm diameter x 1mm width (Figure 1(d)).It has been observed that both grades of magnets have different curie points and interestingly it has been also observed that the size of the magnet has little influence on the Curie point. The possible reason would be the overall magnetization of the magnet is coupled ferromagnetically to the lattices of the transition metal irrespective of the size of the magnet. The Curie temperature of the magnet samples exposed in the sodium environment was also found to be the same as that of unexposed magnet samples.

It can be noted here that the Curie temperatures reported in the literature for N38 and N38SHX magnets are 310 °C and 330 °C respectively [19] which are widely different from the results obtained from this experimental work (255 °C and 270 °C respectively).

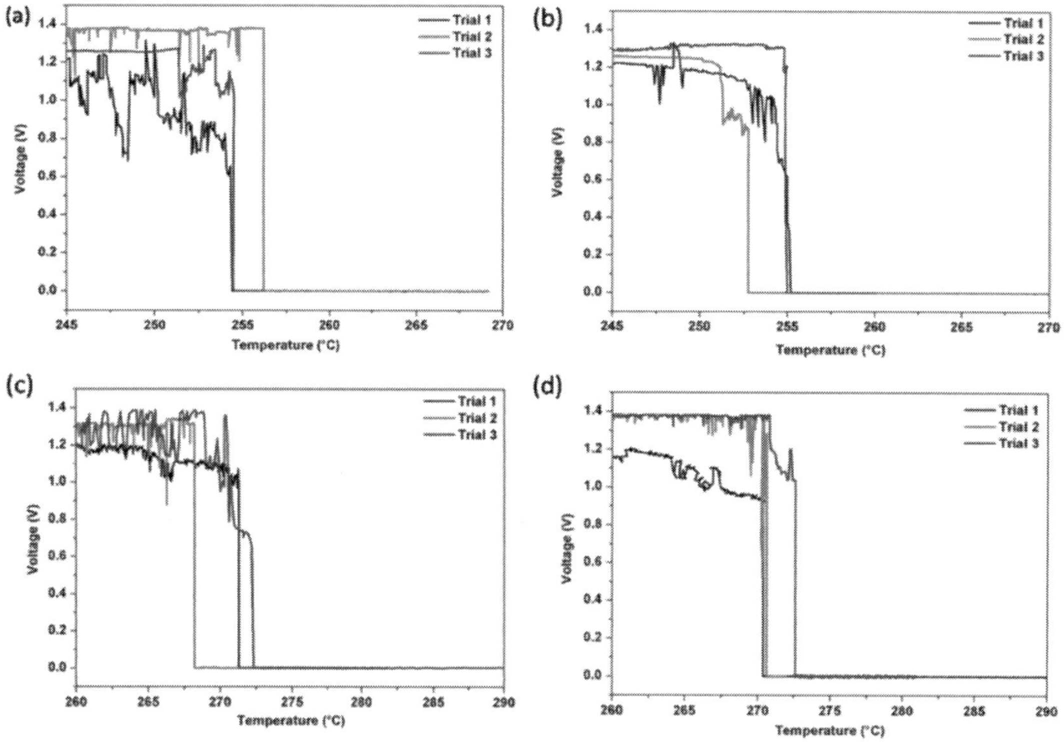

Figure 5: Voltage versus temperature graph for magnet (a) grade N38 size 15mm diameter x 5mm width (b) grade N38 size 5mm diameter x 3mm width (c) grade N38SHX size 10mm x 4mm x 2mm (d) grade N38SHX size 8mm diameter x 1mm width

The current study is towards exploring the feasibility of temperature measurement under reactor operating conditions of specimens during irradiation experiments, where such measurements are complex. This study can further be extended to determine the effect of irradiation on the magnets during the measurement of temperature by the use of a similar experimental setup.

5. Conclusion

This work demonstrates a possible method of using permanent magnets to determine the temperature indirectly during irradiation experiments. The experimental results show that magnets samples selected for the study (N38, N38SHX) are compatible with sodium and it is also found that exposure to sodium environment does not have any effect on Curie temperature of the magnet samples. It was found that the shape and size of the magnet do not affect the Curie temperature of the magnets. The peak temperature can be determined within ±5 °C. The advantages of this method are their small size, easy measurement, no need for encapsulation, and reliability of the indicated temperature to within a few degrees. It is also found that the Curie temperature reported in the literature is approximate since Curie temperature depends on many factors including the chemical composition of the specific magnet chosen. The peak temperature detection with permanent magnets depends on two factors: (1) the temperature profile should be guessed or calculated closely to enable the inclusion of many magnets with different Curie temperature, and (2) the availability of permanent magnets with a small difference in Curie temperature in the selected range of temperature among them. The best measurement of temperature occurs when one of the two adjacent magnets demagnetizes, and the other does not. If the temperature profile cannot be guessed closely, a wide range of magnet with different Curie points should be used. This work can further be extended to determine the effect of irradiation on the magnetic property of the material during the measurement of temperature.

Acknowledgments

The authors gratefully acknowledge the contributions of Mr. Rajesh Saxena and Mr. R. Ramesh, IDEAS, RHIED, CDHG, RDTG, IGCAR during the development, and fabrication of experimental setup.

References

[1]. Baldev R 2009 Materials science research for sodium cooled fast reactors. *Bull. Mater. Sci.* **32** 271–283.
[2]. Srinivasan G, Suresh Kumar K V, Rajendran B and Ramalingam PV 2006 The Fast Breeder Test Reactor-Design and operating experiences. *Nucl Eng Des.* **236** 796–811.
[3]. Kiritani M 1988 the need for improved temperature control during reactor irradiation, *J. Nucl. Mater* **160** 135-141
[4]. Nita N, Yamamoto T, Iwai T, Yasunaga K, Fukumoto K and Matsui H 2002 Effects of temperature change on microstructural evolution in vanadium alloys under ion irradiation up to high damage levels, *J. Nucl. Mater* **307-311** 398–402
[5]. Scervini M, Rae C and Lindley B 2013 Transmutation of thermocouples in thermal and fast nuclear reactors, *3rd International Conference on Advancements in Nuclear Instrumentation, Measurement Methods and their Applications (ANIMMA)*, IEEE, Marseille, France
[6]. Davis K L, Knudson D L, Daw J E, Rempe J L and Palmer A J 2012 Melt wire sensors available to determine peak temperatures in ATR irradiation testing, *Eighth American Nuclear Society International Topical Meeting on Nuclear Plant Instrumentation, Control and Human-Machine Interface Technologies*, INL/CON-11-23913
[7]. Rempe J L, Condie K G, Knudson D L and Snead L L 2010 Silicon Carbide Temperature Measurements at the High Temperature Test Laboratory, *Idaho National Laboratory*, INL/EXT-10-17608
[8]. Cerchione J D and Wilkes C W 1981 Manufacture and use of thermal-expansion-difference temperature monitors, *Trans. Am. Nucl. Soc.* **38** 676-677
[9]. Cullity B D and Graham C D 2009, *Introduction to Magnetic Materials*, Hoboken, New Jersey Wiley pp 477-504
[10]. Stoelzl D and Sizmann R 1972 Magnetic monitors for in-pile measurements of maximum temperature in irradiated samples *J Phys **E** Sci Instrum.* **5** 592-594
[11]. Marth W, Stoelzl D, Sizmann R 1966 Magnetic methods for measuring the maximum temperature of irradiation sample *Kerntechnik* **8** 208-210
[12]. Kim A S 1998 Temperature stable permanent magnet *US Patent 5772796* (to YBM Magnex International, Inc. Newtown, Pa.)
[13]. Hideharu N, Suguru N and Satoru H 2007 Rare earth element permanent magnet material *US Patent 7175718B2* (to Mitsubishi Denki Kabushiki Kaisha, Tokyo, Japan)
[14]. Yutaka M, Masato S and Setsuo F 1986 Process for Producing Permanent Magnet Materials *US Patent 4597938* (Sumitomo Special Metals Co. Ltd., Osaka, Japan)
[15]. Chen C H, Talnagi J, Liu J, Vora P, Higgins A and Liu S 2005 The Effect of Neutron Irradiation on Sm_2Co_{17}-Based High Temperature Magnets and Nd-Fe-B Magnets *IEEE Trans. Magn* **41** 3832 – 3834
[16]. Brown R D, Bush E D and Hunter Jr W T 1982 Radiation Effects on Samarium-Cobalt Permanent Magnets, *Los Alamos National Laboratory*, LA-9437-MS, DE82 019492

IOP Conf. Series: Materials Science and Engineering **992** (2020) 012002 doi:10.1088/1757-899X/992/1/012002

[17]. Sery R S, Lundsten R H and Gordon D I 1961 Radiation Damage Thresholds For Permanent Magnets, *United States Naval Ordnance Laboratory*, White Oak, Maryland AD0262051
[18]. Paulsen J A, Lo C C H, Snyder J E, Ring A P, Jones L L and Jiles D C 2003 Study of the Curie temperature of cobalt ferrite based composites for stress sensor applications *IEEE Trans. Magn* **39** 3316 - 3318
[19]. https://www.arnoldmagnetics.com/wp-content/uploads/2019/06/Arnold-Neo Catalog.pdf , access date: 10/06/2019

IOP Conf. Series: Materials Science and Engineering **992** (2020) 012003 doi:10.1088/1757-899X/992/1/012003

Theoretical and experimental modal analysis of centrifugal pump radial flow impeller

M N Oza[1], Dr D S Shah[2]

[1]C S Patel Institute of Technology, Department of Mechanical Engineering, Faculty of Technology and Engineering, Charotar University of Science and Technology, Anand – 388421, Gujarat, India

[2]Mechanical Engineering Department, Sardar Vallabhbhai Patel Institute of Technology, Anand, Gujarat, India

E-mail: madhavoza.me@charusat.ac.in

Abstract. Modal analysis is extensively used to identify the dynamic properties of a structure in terms of the dynamic characteristics: natural frequency, damping factor and mode shape.In this research, modal analysis has been carried out on a radial flow impeller of a centrifugal pump. The 3-D model of the impeller was built in CAD software Creo Parametric 1.0 from a 2-D drawing and that model was imported in FEA software package ANSYS for conducting modal analysis. The model was meshed effectively using SOLID187 elements. The first twenty natural frequency and mode shapes were extracted and nodal diameter was identified for each mode shape. The experimental modal analysis was conducted using impact hammer and the first twenty natural frequencies were identified using FFT analyzer. The natural frequencies of both the methods were compared and it was found that there is a good agreement between these results, thus the FE model is validated. The results are discussed and conclusions drawn.

1. Introduction

A centrifugal pump radial impeller is the rotating part transforming driver energy of the fluid into the kinetic energy. Due to the uncertain pattern of flow generated by the volute inside the radial impeller and whirling of impeller shaft, many dynamic forces are developed, leading to vibrations in the pump. The fatigue failure of the impeller is caused by fluctuating forces in combination with steady forces. The blades of the impeller are very critical components and they are exposed to very turbulent conditions. The blades of an impeller are frequently under very hostile operating conditions and damage the machine [1]. Reliability of impellers is very important parameter affecting the effective operation of the pump as well as environmental safety [2]. To design reliable impellers, the blade and disk modal analyses are of the utmost important. The evaluation of an impeller design is made by combining the dynamic behaviour with the nature of the fluctuating forces. Greater knowledge of the vibration behaviour of radial-flow impellers will be useful for optimizing their fatigue-life and efficiency, possibly resulting in increased performance and applicability. Modal analysis is the method to identify the dynamic characteristics of any system in terms of natural frequencies, mode shapes and damping factor to developed a mathematical model to predict its dynamic behavior [3]. Ramana Podugu et al. conducted modal analysis for condition monitoring of pump vibration and extracted natural frequency for the pump [4]. Modal analysis is generally performed using finite element method and then the results are validated by conducting modal testing as finite element model gives approximate results to any numerical problem. L. Subramaniam and S. Sendilvelan performed modal analysis to identify the dynamic characteristics of centrifugal pump impeller with different blade thickness [5]. Samir Lemes et al. investigate the mode shapes of centrifugal impeller by comparing them with theory of circular platen and extracted Chladni figures identifying nodal diameter and nodal circles [6]. Mode shapes of impeller are also necessary for acoustic fatigue analysis and resonance identification of rotating machineries [7-10].

The modal analysis of the centrifugal pump impeller is performed using FEA software package ANSYS 14.0 and natural frequencies and mode shapes are extracted. Nodal diameter pattern is found in the mode shapes and up to five nodal diameter have been found in the mode shapes. To validate the results of the ANSYS, the experimental modal analysis has been performed using impact hammer and the results of both the methods are compared. The good agreement was found in results of both the methods.

2. Modal Analysis of Impeller using FEA

FEA modal analysis of the impeller was carried out for the identification of natural frequencies and mode shapes using FEA package ANSYS 14.0. The impeller is of open radial type having five curved blades. First of all, the 3D model of the impeller was developed in CAD software Creo Parametric 1.0 as shown in figure 1 and imported in ANSYS workbench for modal analysis. Total twenty natural frequency and mode shapes were extracted and nodal diameter for each mode was identified using visual inspection. To carry out modal analysis in ANSYS Workbench following steps were performed.

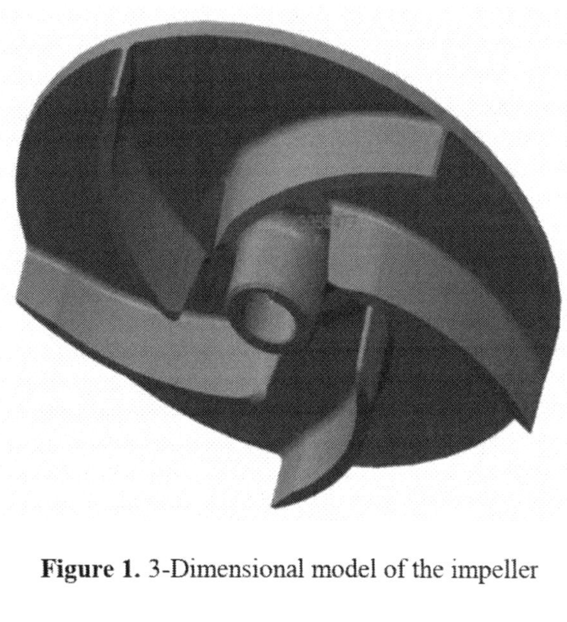

Figure 1. 3-Dimensional model of the impeller

2.1 Specifying material properties

The impeller is a casted part of the material AISI 316L. The material properties used in analysis are: Modulus of Elasticity = 190 GPa, Density = 7750 kg/m^3, Poisson's ratio = 0.25.

IOP Conf. Series: Materials Science and Engineering **992** (2020) 012003 doi:10.1088/1757-899X/992/1/012003

2.2 Meshing and boundary conditions

The impeller is composed of a complex geometric structure, and hence, for better results a higher order tetrahedral element SOLID187 is used. This element is a 10-node tetrahedral element [11]. The information of the SOLID187 element is shown in figure 2. The total no of nodes and elements are 69920 and 39943 respectively. The 3-D model of impeller very accurately meshed such that it exhibits the accurate behaviour for modal analysis. After effective meshing of the impeller, the boundary conditions were applied. The impeller was constrained by applying a cylindrical support at its centre. The meshed model of the impeller is shown in figure 3.

2.3 Solution and results

First twenty natural frequency and modes were extracted in the ANSYS Workbench using the Block Lanczos method of mode extraction. The natural frequencies extracted are plotted in table 1 and figures 4 to 9 shows the corresponding mode shapes. As discussed in previous section the results extracted for modal analysis of the impeller are mode shapes and corresponding natural frequencies. Total twenty mode shapes and natural frequencies were obtained from ANSYS Workbench solver of the impeller. The mode shape is the deformation pattern of the impeller corresponding to the particular natural frequency. The mode shapes and natural frequencies were saved in result directory for further analysis.

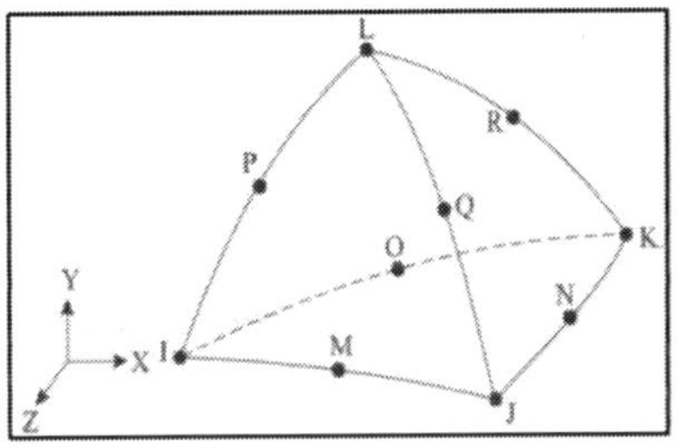

Figure 2. Spatial element Solid187

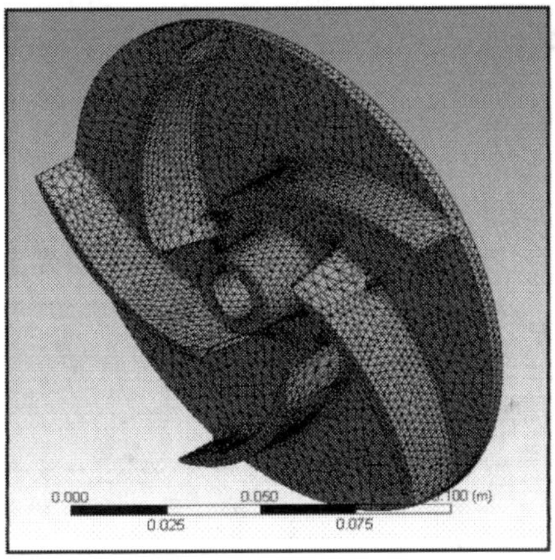

Figure 3. Meshed model of the impeller

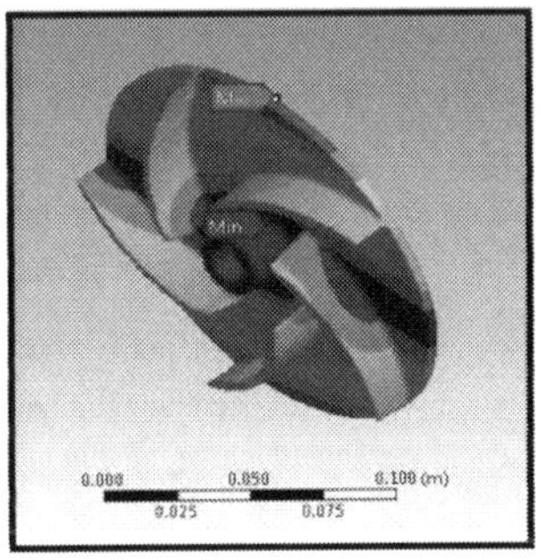

Figure 4. 1st mode shape

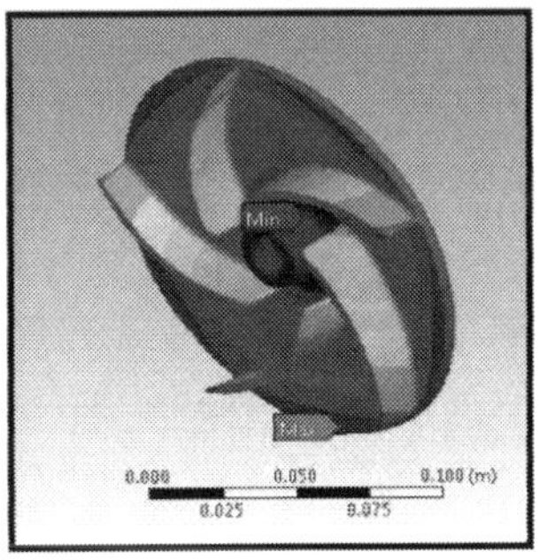

Figure 5. 3rd mode shape

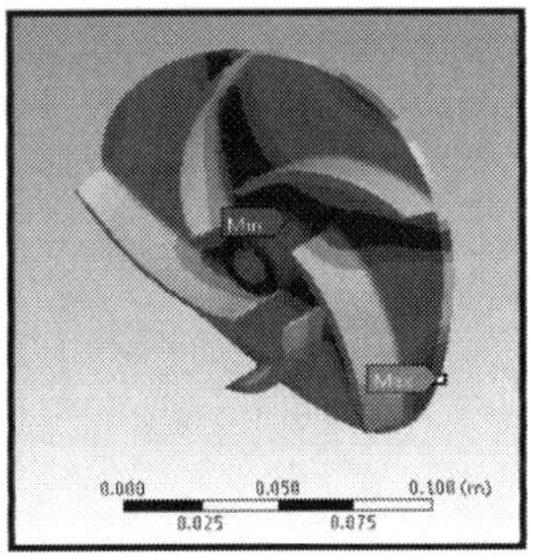

Figure 6. 5th mode shape

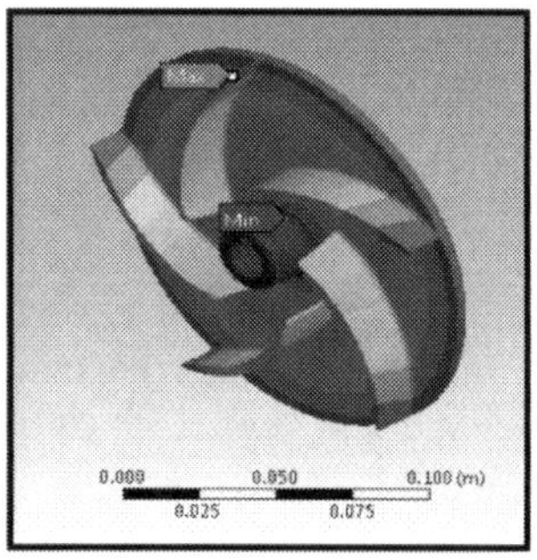

Figure 7. 6th mode shape

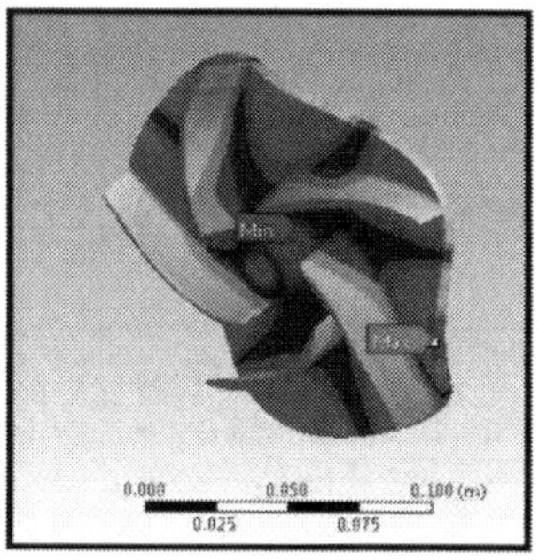

Figure 8. 10th mode shape

Figure 9. 12th mode shape

After closely inspecting each mode shapes and their corresponding natural frequencies of the impeller, some observations have been identified. The first natural frequency is 631.1Hz which is very high for the centrifugal pump impeller. Due to very close eigenvalues the repetitive natural frequencies are found, which is very clear from table 1. As shown in table that first natural frequency is 631.1Hz while second natural frequency is 633.4Hz and the highest natural frequency extracted is 7615 Hz.

The mode shapes of the impeller are combined disc and blade mode shapes. For an open bladed impeller, the mode shape is either characterized by circular nodal lines or nodal diameter patterns, or combinations of both. The nodal diameter pattern has been identified in the mode shapes of the impeller. Generally nodal diameter pattern are being seen in pairs, which means that two consecutive mode shapes have same nodal diameter. The same observations have been identified in the mode shapes of the impeller, which is shown in table 1 below. While in the modes starting from eleven up to twenty the modes are combined and so very complex nodal diameter pattern were found.

Table 1. Natural frequencies, modes and Nodal diameter for the impeller

Mode	Natural Frequency (Hz)	Nodal Diameter	Mode	Natural Frequency (Hz)	Nodal Diameter
1	631.1	1	11	4994.84	-
2	633.4	1	12	5403	-
3	825.29	0	13	5552.77	-
4	1051.6	2	14	5562.94	-
5	1051.9	2	15	5921.96	-
6	1832	0	16	6414.12	-
7	2255.7	3	17	6420.41	-
8	2256	3	18	6620.47	-
9	3720.4	4	19	6622.80	-
10	3721.1	4	20	7615.60	-

The first and second mode shape exhibits one nodal diameter pattern. There are two phase change in the impeller which is noted by the radial lines forming the nodal diameter mode shape. Third and Sixth mode shape forms zero nodal diameter pattern, which is circular mode (umbrella mode) where the

IOP Conf. Series: Materials Science and Engineering **992** (2020) 012003 doi:10.1088/1757-899X/992/1/012003

maximum deflection is at the blade tip and outer periphery of the disc of the impeller. Fourth and Fifth mode shapes form second nodal diameter pattern while and seventh and eighth mode shape form three nodal diameter pattern. In all these mode shapes there is maximum deflection is seen at the outer periphery of the impeller. From first to tenth mode shape of the impeller there is not much blade deflection is seen except third and sixth mode shape. From eleventh to twentieth mode shapes the high blade deflection is seen. Also in these mode shapes the deflection in the blades occurs near the center of the impeller. Nodal diameter pattern in the mode shapes is identified which can be seen from figure 4 to figure 9.

3. Experimental Modal Analysis of the Impeller

The experimental modal analysis is conducted to validate the natural frequencies obtained from the finite element analysis. For doing the experimental modal testing, the impeller was fabricated by casting. The method used to excite the impeller was impact hammer testing. The process of modal testing was divided in three phases: 1.Pre-test Planning 2.Development of setup 3.Actual test and extraction of results. In pre-test planning the method of placing the impeller was decided and the location of points of excitation and response measurements were identified. In second phase the test setup was developed for placing the impeller and in final stage the test was performed and results were extracted using FFT analyzer and analyzed in a computer.

The centrifugal pump impeller is tested for extracting natural frequency using modal testing method. The test setup was developed by fabricating a steel frame where a shaft is placed on which impeller is fixed. The test setup developed for modal testing is shown in figure. The selection of points of excitation and response measurements were identified from the FE model developed earlier. The impeller was fixed on a shaft placed on a fabricated steel frame. The shaft was turned on a lathe machine with the dimension of 14.8 mm diameter such that the impeller was fixed on it in all directions.

After development of the test setup, the selection of points of excitation and response measurement is decided. The points of excitation were identified on the blade of the impeller which is the most critical part of the impeller. The points selected for response measurements were decided on two conditions. The first condition is that the points near to any nodal line of the mode shape were avoided and the second condition is to selecting the points with extremely low vibration amplitude. After visually inspecting each mode shapes developed in FEA the location of points of excitation and points of response measurements were identified on the blade of impeller.

The instruments used in modal testing of the centrifugal pump impeller were impact hammer for the force excitation, accelerometers for response measurements, vibrometer for calculating FRFs and a computer to process the FRFs and generate frequency spectrums. The accelerometers used were piezoelectric type and the vibrometer was a 2-channel vibration analyzer. To analyze the frequency response extracted by the vibrometer, the computer system was used. An impulse hammer consisting of a transducer excited the impeller structure and the input load was measured. The output response of the impeller vibration was measured using an accelerometer attached to the impeller. The signals, which were generated as a result of the above method, were processed by the FFT analyzer connected to a computer system. The computer system coverts the digital frequency domain data into frequency response functions. After analyzing the FRF data the natural frequencies were identified. The natural frequencies were extracted from the frequency window of 0-10000Hz and 0-8000Hz of FFT analyzer. The frequencies extracted from the vibrometer are shown in figure 9 and 10. The peaks in the figure 9 and 10 exhibit the excited natural frequencies of the centrifugal pump impeller which are 1090 Hz, 2355 Hz, 3688 Hz to 7598 Hz.

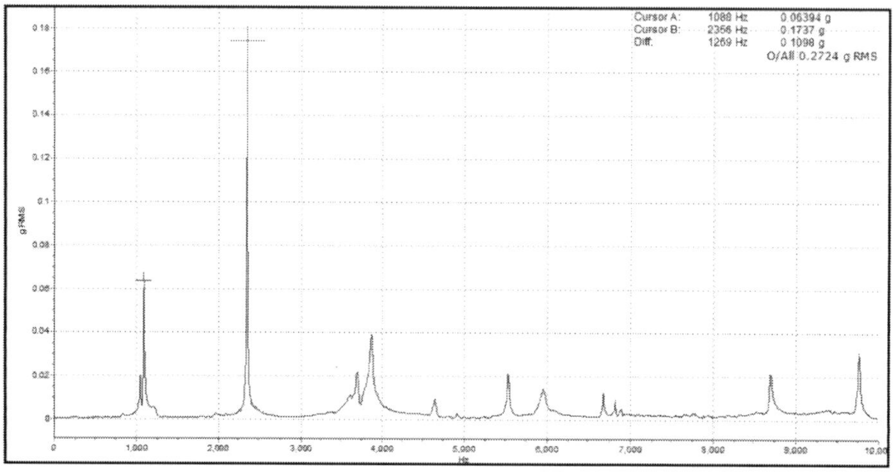

Figure 10. Natural frequency spectrum extracted for the impeller in the range of 0-10000Hz

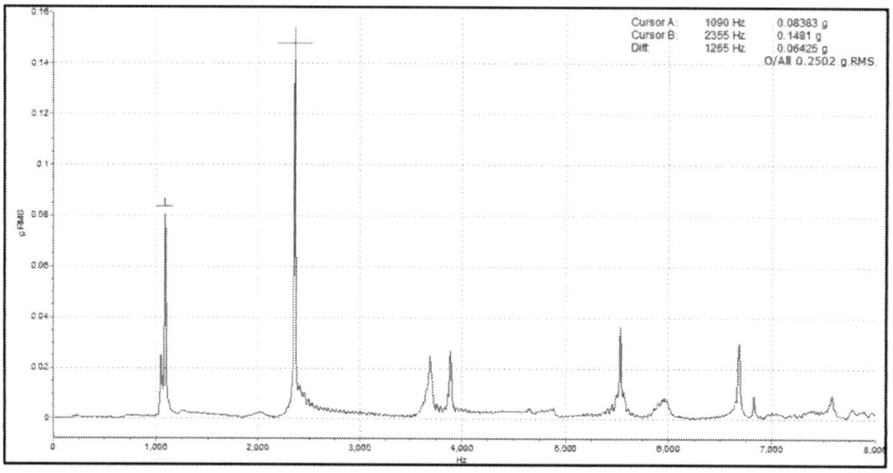

Figure 11. Natural frequency spectrum extracted for the impeller in the range of 0-10000Hz

Table 2. Natural frequency extracted in experimental modal analysis

Sr No.	Frequency
1	1090.00
2	2355.00
3	3688.42
4	4650.91
5	5532.25
6	5963.25
7	6677.89
8	6821.05
9	7598.94

IOP Conf. Series: Materials Science and Engineering **992** (2020) 012003 doi:10.1088/1757-899X/992/1/012003

The main objective of the research work is to validate the finite element model of the impeller by comparing the results of modal analysis using ANSYS Workbench and Experimental modal analysis. The results have been extracted by both the methods effectively. The 3-D CAD model used for the analysis in ANSYS Workbench was the CAD model of the original casted component used in modal testing. The boundary condition was also applied in the ANSYS Workbench such that it can match with the experimental setup developed for testing. So the method used to correlate the results of both the techniques of modal analysis is to compare the natural frequencies extracted of the impeller. There are some frequencies which either not excited in ANSYS Workbench or in experimental method. So only those frequencies are selected for comparisons which were excited in both the methods. The comparison of the natural frequencies is as shown in table 4.

Table 3. Comparison of natural frequencies of ANSYS Workbench and experimental modal analysis of the impeller

Natural Frequency (Hz)		**% difference**
ANSYS Workbench	Experimental Modal Analysis	
1051	1091	3.805899
1832	N.A.	N.A.
2256	2355	4.388298
3720	3688.42	-0.84892
4994	4650.91	-6.87004
5403	5532.25	2.39219
5562	5963.25	7.214132
6414	6677.89	4.114281
6620	6821.05	3.037009
7615	7598.94	-0.2109

4. Conclusion

Modal analysis was performed to calculate natural frequencies of centrifugal pump impeller. The natural frequencies were first extracted by conducting modal analysis using finite element method in ANSYS 14.0 software. The three dimensional model was developed in Creo parametric 1.0 software and it was imported in ANSYS Workbench to extract first twenty natural frequencies and mode shapes. The modal testing was performed using impact hammer and accelerometer on the impeller to generate to validate the results of the FEA method. The response was calculated using FFT analyzer. The natural frequencies of impeller were compared for both of the methods and good correlation was found. The maximum percentage deviation of the results of modal analysis using FEA and modal testing was seven percentage. Hence the modelling in ANSYS software was found to be validated for dynamic analysis.

References

[1] Ziaei R S 2005 *Iranian J. of Sci. & Tech. Trans. B Engg.* Finite element, modal testing and modal analysis of a radial flow impeller **29 B2** 157-169

[2] Schwarz B J and Richardson M H 1999 *CSI Reliability Week* Experimental modal analysis Orlando, FL

[3] He J and Fu Z 2000 *Modal Analysis* 1stEdn Butterworth-Heinemann 2-9

[4] Podugu R, Kumar S J, Murthy B V R and Kumar N S 2011 *Int. J.l of Engg. Sci and Tech.* A modal approach for vibration analysis and condition monitoring of a centrifugal pump **38** 6335-6344.

[5] Subramaniam L and Sendilvelan S 2012 *European J. of Scientific Research* Modal analysis of a centrifugal pump impeller **79 1** 5-14

[6] Lemeš S and Zaimović Uzunović N 2002 *6th International Research/Expert Conference Trends in the Development of Machinery and Associated Technology* Mode shapes of centrifugal pump impeller Neum B&H 18-22.

[7] Eisinger F. L 2002 *Trans. of ASME, J. of Pressure Vessel Tech.* Acoustic fatigue of impellers of rotating machinery **124**

[8] Singh M P, Thakur B K, Sullivan W E and Donald G 2003 *Proc. of the Twenty-Eight Turbomachinery Symp.* Resonance identification of impellers Texas A&M University College Station TX

[9] Wang Q, Bartos J C and Houston R A 1999 *Proc. of the 28th Turbomachinery Symp.* Methodology of open bladed impeller resonance identification Texas A&M University College Station TX USA

[10] Kushner F, Richard S J and Strickland R A 2000 Proceedings of the Twenty-Eight Turbomachinery Symposium Critical review of compressor impeller vibration parameters for failure prevention Texas A&M University College Station TX

[11] ANSYS Workbench 14.0 Help

IOP Conf. Series: Materials Science and Engineering **992** (2020) 012004 doi:10.1088/1757-899X/992/1/012004

Design optimization of magnetorheological brake using structural parameter: evaluation and validation

S Patel[1*], R Upadhyay[1] and D Patel[2]

[1] P. D. Patel Institute of Applied Sciences, Charotar University of Science and Technology (CHARUSAT), Changa-388421, Gujarat, India

[2] C. S. Patel Institute of Technology, Charotar University of Science and Technology (CHARUSAT), Changa-388421, Gujarat, India

*E-mail: 15drnst001@charusat.edu.in

Abstract. Magneto-rheological (MR) brake is one of the most promising smart systems, which transmit torque with the applied magnetic field. This process is reversible in the absence of magnetic field. In this paper, the most common disk type MR-Brake (MRB) design is selected to optimize for the desired range of torque transmission without changing the diameter of the MRB. The theoretical torque transmission equation is deduced and using that power index dependency is studied for each structural MRB parameter. Three structural parameters (outer radius of the disk, fluid gap, and coil width) are selected, for three levels and analysis has performed using Minitab software to understand percentage contribution of the structural parameters. Then by varying fluid gap and fixing the other two MRB parameters, FEMM analysis has performed to understand magnetic flux concentration at the fluid gap. After confirming the MRB design, MRB is developed having two different MR-fluid working gaps. Later, maximum torque transmission of the MRB having different MR-fluid working gaps are studied at applied current value. The experimental data confirms with theoretical calculation.

1. Introduction

Magnetorheological (MR) fluid is considered in the category of smart material due to its phase change (fluid to quasi-solid) process in the presence of a magnetic field (MF). This process is reversible when MF is removed and the time required is in milliseconds. The MF tune the rheological properties over three to four orders of magnitude. The MF induces a yield-stress in fluid; this property is used in many applications [1-5]. One such application is rotary MR-Brake (MRB). The MRB transmits torque by the shear resistance of an MR-fluid. In developing MRB, important aspects that should be considered carefully are size and braking torque. Using the different configurations of MRB braking torque can be enhanced. The disk-type MRB is the most common configuration, which is also the pioneer configuration described by Jacob Rabinow in 1951 [6]. Nevertheless, for development and improvement of MRB performances several quality research work has been done [7-9]. These research works substantial efforts are made to improve brake design.

The primary input of this study is to optimize the rotary-disk MRB for practical application and validate the same. The developed MRB design is selected from available literature and the targeted application is for mid-sized motorcycle [10]. To optimize the same the comprehensive suite of data analysis and process development tools were used. The percentagewise contribution of the effective structural parameter is determined by the Minitab software [11]. FEMM [12] analysis performed to understand magnetic flux concentration and leakage in the MRB design, which facilitated a lot in choosing the MRB material right. Grounding on these, MRB is developed and its performance is evaluated.

2. MRB Theory

The rotary disk type MRB involves three components for total torque (T) transmission. These component followed as, (i) fluid viscosity, (ii) friction, and (iii) field-dependent yield stress [13]. Therefore, total torque is represented by,

$$T = T_{viscosity} + T_{friction} + T_{yield-stress} \tag{1}$$

The Herschel-Buckley model is used for MR-fluid under applied MF condition for high shear rate applications. It is represented with shear stress (τ), in the MRBs it can expressed as [14],

$$\tau = \tau_{yd} + K\dot{\gamma}^n \tag{2}$$

τ_{yd} is yield stress of the MR-fluid, $\dot{\gamma}$ is shear rate, n and K are flow behaviour index and consistency parameter respectively. Researchers [15, 16] have given the mathematical expression for the torque transmission of the single disk rotary MRB based on equation (1) and (2). The mathematical form for the same is expressed as equation (3).

$$T_{total} = 2\pi R_{OD}^2 b_D K_0 \left(\frac{\omega R_{OD}}{h_{duct}}\right)^{n_0} + \frac{4\pi\tau_y}{3}\left(R_{OD}^3 - R_{ID}^3\right) + \frac{4\pi\mu_{eq} R_{OD}^4 \omega}{(n+3)h}\left[1 - \left(\frac{R_{ID}}{R_{OD}}\right)^{n+3}\right] \tag{3}$$

Where, R_{ID}, R_{OD}= inner and outer disk radii respectively, ω= rotary disk angular velocity, h= fluid gap, h_{duct}= gap of the annular duct, b_D= thickness of the disk, n_0, K_0= flow behaviour index and constancy at zero MF, μ_{eq}= equivalent viscosity. Each brake parameter's effect on torque has been estimated. The power index dependency of the MRB parameters are shown in figure 1a. The torque of the MRB is majorly governed by two parameters. First is the τ_{yd} and second is the R_{OD} of the MRB. These two parameter dominates the other parameter. So, the performance of the MRB can be improved by enhancing yield stress of MR-fluid and optimizing outer radius of the disk. The yield stress of MR-fluid is the function of the MF intensity (Figure 1b). So, to enhance MR-fluid yield stress, MF intensity can be optimized without changing fluid property. However, the MF intensity explicitly depends on coil width and MR-fluid working gap. So, keeping this in mind, in next section MRB design, its effective parameters, and its optimization are discussed.

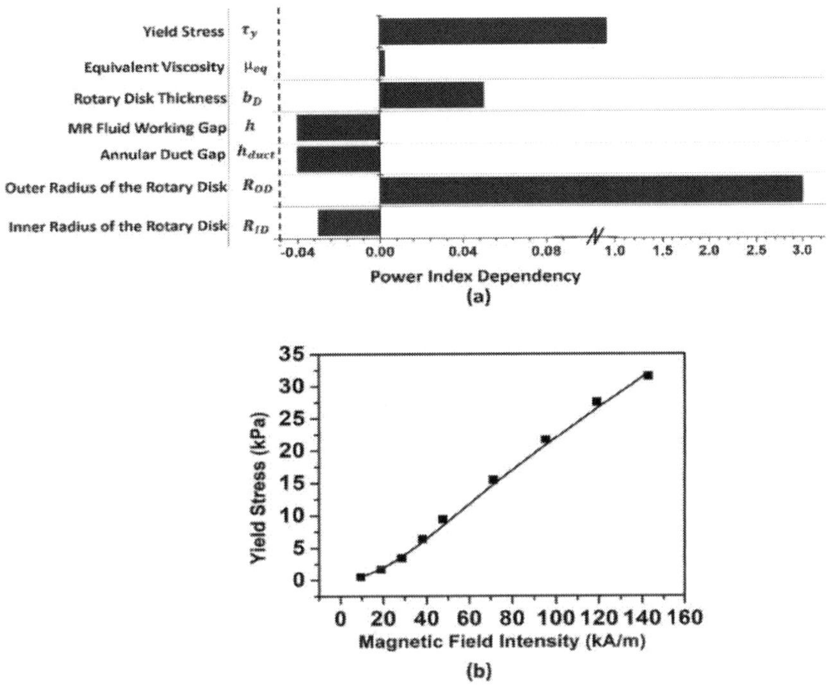

Figure 1. (a) Power index dependency of MRB parameter on torque transmission (b) The yield stress variation with MF intensity for MR-fluid. The line is fit to proposed Fang et al. [17] empirical relation of yield stress versus field.

3. MRB Design

In MRB, the electromagnetic coil (EM-coil) encloses the rotary disk in the central casing. MR-fluid region is energized by EM-coil. Each side is bolted with a side casing. MR-fluid is occupied in the space formed by the side casing and rotary disk. Figure 2 depicts the cross-sectional view of the MRB. Table 1 shows the MRB parameters (except R_{ID}, R_{OD}, h, b_D).

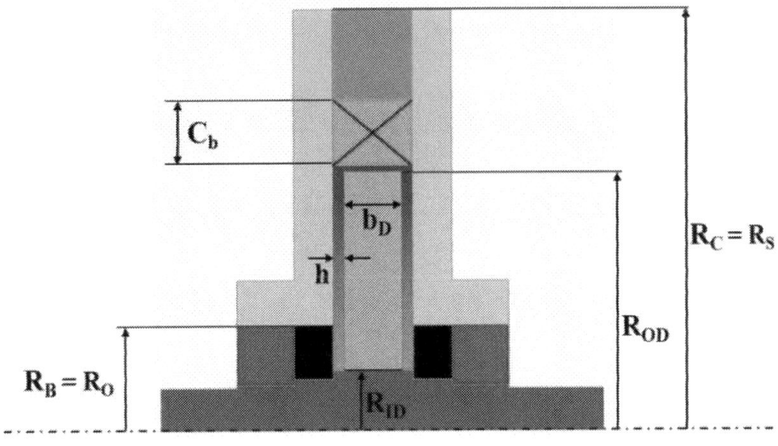

Figure 2. Cross-sectional view of the MRB with its structural parameter.

Table 1. MRB parameter

Sr. No.	Parameter Symbol	Parameter
1	R_B	Radius of Bearing
2	R_O	Radius of oil-seal
3	R_C	Central casing radius
4	R_S	Side casing radius
5	C_b	EM-coil width

As discussed in the earlier section, MRB performance can be enhanced by improving fluid yield stress and optimizing essential structural parameters. By keeping the aforementioned optimization parameter of MRB, brake performance is optimized using FEMM (Finite Element Method Magnetics) software considering parameter as in table 2. There are seven basic structural parameters (bearings and oil seal dimensions are not considered) of an MRB which contributes to the performance of the torque transmission. Grounded on power index dependency, three structural parameters are selected for optimization which dominates the torque transmission among all. The selected parameters are the outer radius of rotating disk (R_{OD}), EM-coil width (C_b) and MR-fluid working gap (h). By using these three parameters in Minitab software, the full factorial design of experiment table 3 is prepared. In that outer radius, coil width and MR-fluid gap are input parameters, and braking torque and power requirement (current supplied to EM-coil) are considered as output parameters. For this analysis, the current supply to the magnetic coil is 1 A so, power is calculated based on the resistance of the coil using equation $P=I^2R$.

The analysis is performed (EM-coil current=1.0 A) to find out the torque and power in FEMM software for different combinations of radius, coil width, and gap. The values of torque and power are shown in table 3. For statistical analysis, Minitab software is used. To find, effect of outer radius of disk, coil width and MR-fluid gap on the torque and power main effect plot is used. By using the main effect plot as shown in figure 3a the torque increases with coil width and disk radius. As higher coil width produces higher MF which results in enhanced τ_{yd} and based on equation (3) higher the radius maximum torque will be produced. However, as the MR-fluid working gap increases the toque decreases due to decrement in the magnetic flux which leads to lower τ_{yd}. Figure 3b shows the main effect plot for power. The plot shows that with increase in radius and coil width power requirement increases while with the gap it is unchanged. With the change in the coil width and radius, the number of turns wound in the coil changes. Nevertheless, coil configuration is independent of the MR-fluid gap, thus power requirement remains constant.

Table 2. Effective MRB parameters for optimization

Sr. No.	Parameter Symbol	Parameter	Lower Range (1-level)	Mid Range (2-level)	Upper Range (3-level)
			Dimensions (mm)		
1	R_{OD}	Outer radius of rotating disk	40	45	50
2	C_b	EM-coil width	8	11	14
3	h	MR-fluid working gap	1	1.5	2

Table 3. Structural Parameters of MRB and its response

Sr. No.	R_{OD} (mm)	C_b (mm)	h (mm)	T (Nm)	R (Ω)
1	50	14	2	4.89	32.8
2	50	14	1.5	5.38	32.8
3	50	14	1	5.80	32.8
4	50	11	2	4.27	23.9
5	50	11	1.5	5.39	23.9
6	50	11	1	6.60	23.9
7	50	8	2	2.94	17.5
8	50	8	1.5	4.10	17.5
9	50	8	1	5.69	17.5
10	45	14	2	4.22	30
11	45	14	1.5	5.23	30
12	45	14	1	5.81	30
13	45	11	2	3.15	21.8
14	45	11	1.5	4.11	21.8
15	45	11	1	5.34	21.8
16	45	8	2	2.32	15.9
17	45	8	1.5	3.07	15.9
18	45	8	1	4.38	15.9
19	40	14	2	3.03	27.2
20	40	14	1.5	3.88	27.2
21	40	14	1	4.54	27.2
22	40	11	2	2.25	19.8
23	40	11	1.5	2.97	19.8
24	40	11	1	4.07	19.8
25	40	8	2	1.65	14.3
26	40	8	1.5	2.20	14.3
27	40	8	1	3.20	14.3

Analysis of variance (ANOVA) is executed to find the contribution of these three parameters on torque and power. Table 4 shows the percentage contribution of each parameter for torque. Where DF represents the amount of information in a data point. SS shows the sum of the square and the adjusted sum of a square is represent as Adj SS. Adj MS shows the adjusted mean square for the given data point. F value shows the response of the input parameters on torque. P values show a significant level of input parameters. In table 4 P-value is less than 0.05 so all the input parameters are considered as significant for the torque. The percentage contribution is calculated from the Adj MS values. The R^2 value is 95.06%.

From table 4, the maximum contribution in the torque is 39.27% for radius followed by a gap of the fluid and coil width. The percentage contribution for the gap of the MR-fluid is 36.62% and the coil width is 23.58%.

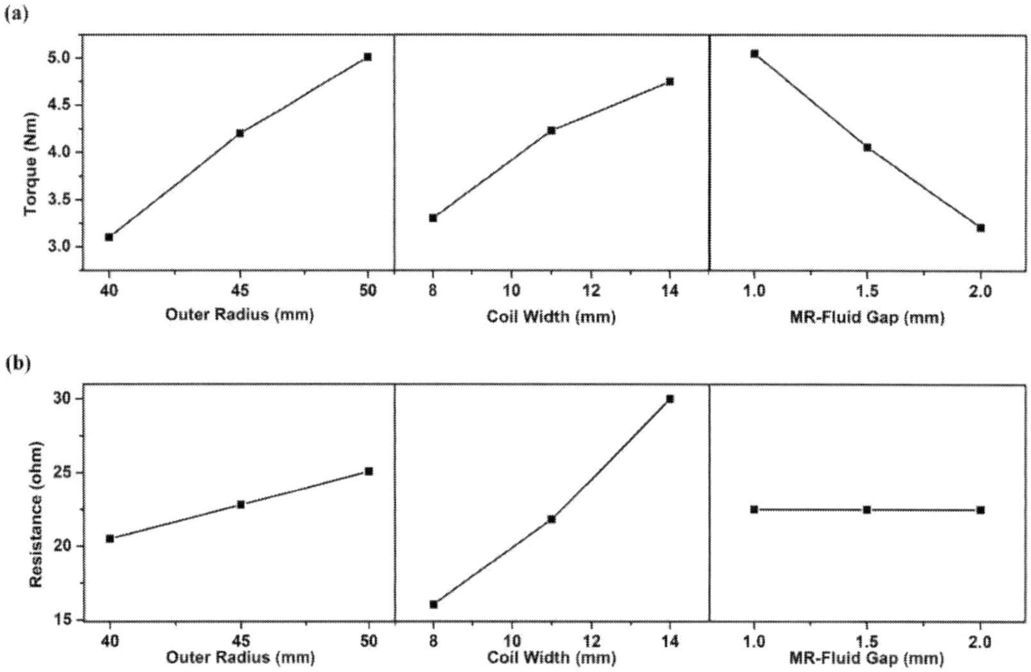

Figure 3. Main effects plot of structural parameter on torque and resistance.

Table 4. Percentage contribution of MRB parameter on Torque

Source	DF	Seq SS	Adj SS	Adj MS	F	P	Contribution
Outer radius	2	16.6772	16.6772	8.3386	76.02	0	39.27
Coil Width	2	10.0159	10.0159	5.008	45.66	0	23.58
MR-fluid gap	2	15.5536	15.5536	7.7768	70.9	0	36.62
Error	20	2.1938	2.1938	0.1097			0.005
Total	26	44.4404		21.2331			

Table 5. Percentage contribution of MRB parameter on Power

Source	DF	Seq SS	Adj SS	Adj MS	F	P	Contribution
Outer radius	2	83.21	83.21	41.6	188.53	0	8.44
Coil width	2	902.13	902.13	451.06	2044.09	0	91.51
MR-fluid gap	2	0	0	0	0	1	
Error	20	4.41	4.41	0.22			
Total	26	989.75		492.88			

Similarly, the percentage contribution in power is due to coil width, which is a maximum of up to 91.51%. The percentage contribution of the radius is nearly 8.44% (table 5), which is very less compared to the coil width. The gap is not a significant parameter, so it does not contribute to power. The R^2 value for the power is 99.55%. From the aforementioned study and analysis, an MRB design was selected keeping mid-sized motorcycles brake parameters. In the next section, design parameters and test results are discussed.

4. MRB test setup and evaluation

To validate the optimized parameters, two sets of MRB designs are considered by varying the MR-fluid working gaps (h=1mm and 2mm). MR-fluid working gap has no effect on power but contributes 37% on torque. The outer radius is not varied as it is fixed for mid-sized motorcycle. So, the outer radius of the disk (R_{OD}=45mm) and coil width (C_b=11mm) kept the same. The detailed MRB parameter values are mentioned in table 6.

Table 6. Dimensions of the fabricated MRB

Sr. No.	Component	Values (mm)	
		MRB-I	MRB-II
1	Rotary disk- inner radius	10	10
2	Rotary disk- outer radius	45	45
3	Rotating disk thickness	12	10
4	MR-fluid working gap	1	2
5	Radius of Bearing	18.5	18.5
6	Radius of oil-seal	18.5	18.5
7	Central casing radius	85	85
8	Side casing radius	85	85
9	EM-coil width	11	11

The MF intensity distribution attained from FEMM for MRB-I and MRB-II is shown in figure 4. The analysis shows that the flux lines remain uniform in the MR-fluid working gap. As the MR-fluid working gap decreases the MF intensity increases. The increased field intensity leads to the robust yield stress of MR- fluid.

The fabricated components of the MRB and MRB assembly are shown in figure 5a and 5b separately. The MRB test rig schematic diagram (figure 6) which contains 2hp AC motor coupled with a gearbox. The MRB is connected to the torque sensor (rotary type), supported by rod-end bearings and jaw couplers. The EM-coil in an MRB is controlled by the DC power supply (30V and 5A).

The performance of the MRB-I and MRB-II has been examined for synthesized MR-fluid (for MR-fluid synthesis, refer [18]). The braking torque variation with the applied current at 300 rpm is shown in figure 7 for MRB-I and MRB-II. Torque value increases with the application of current, the response to applied current is different for MRB-I compared to MRB-II. The line is a theoretical fit to the experimental data using MRB braking torque equation (equation (3)). It is to be noted here that MRB-I has a smaller MR-fluid working gap and produces higher torque values than the MRB-II. This is due to the higher MF intensity concentration at MR-fluid working gap (figure 4). In MRB-I, shaft speed tends to zero after the applied current value reaches 0.6A while for MRB-II it is 0.8A. The torque value in MRB-I is increased by 91.67% compare to MRB-II at 0.6A current. The value of torque obtained for MRB-I and

MRB-II, from equation (3), using τ_y versus H curve at 1 A agrees well with optimized values given in table 3.

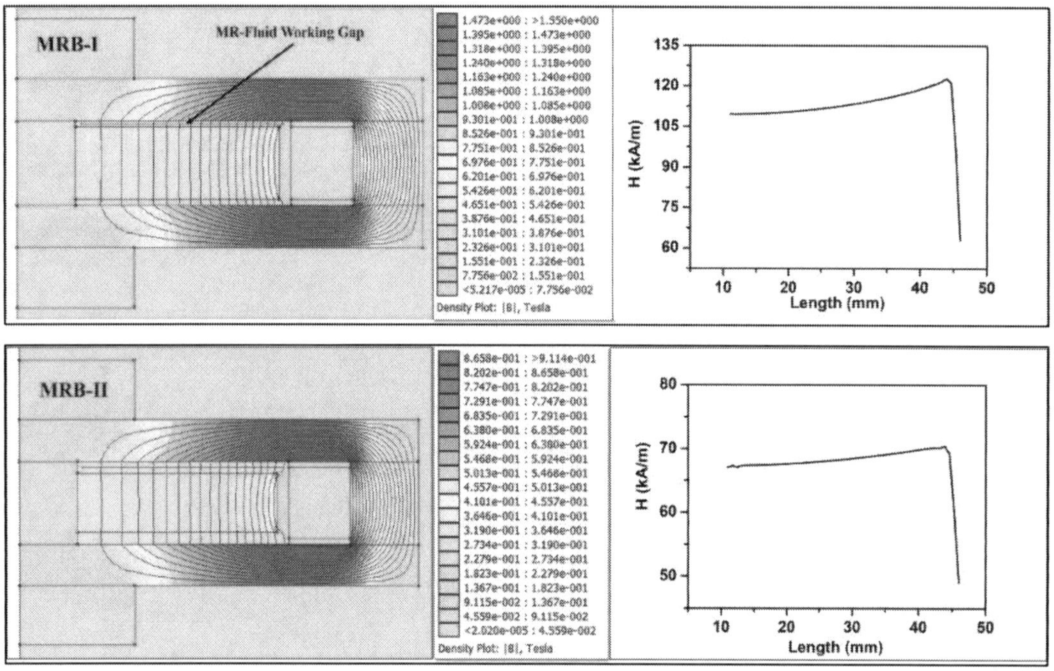

Figure 4. MF distribution of MRB using FEMM (left side) and variation of MF intensity with MR-fluid length (right side). Field remains uniform in the MR-fluid working gap and decreases at the annular duct.

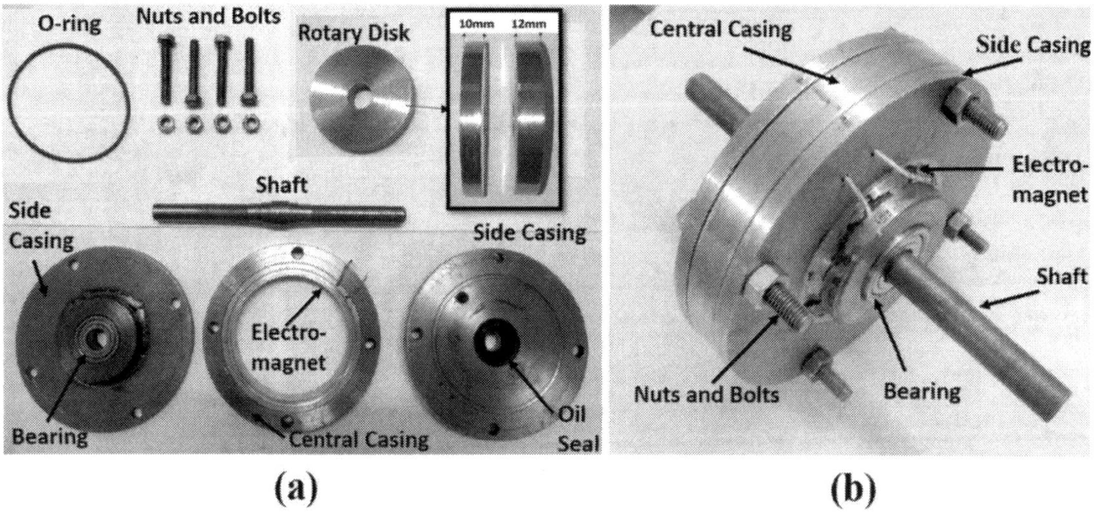

Figure 5. (a) MRB components, (b) Assembled MRB

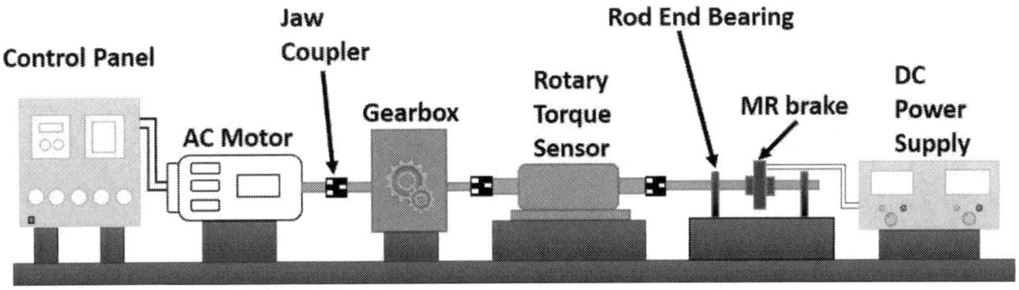

Figure 6. MRB test setup (Schematic Diagram)

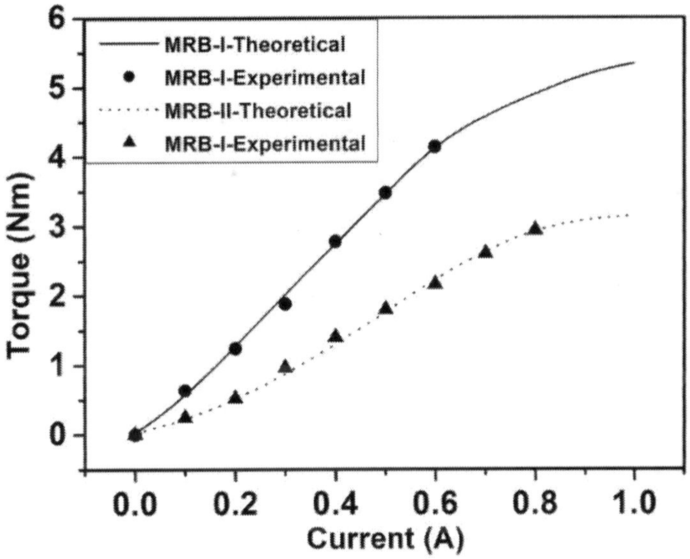

Figure 7. Variation of braking torque with applied current for MRB-I and MRB-II. The line is calculated using equation (3) and τ_y value obtained from figure 1b at different value of current (or MF intensity).

5. Conclusion

The effect of the structural parameters of MRB on torque transmission and its optimization is studied in the present work. Power index dependency of MRB parameters was obtained, which helped to zero down the most effective structural parameters of MRB. From this, three parameters were selected for three different levels and analyses were performed using Minitab software. ANOVA was carried out to find out the contribution of three parameters on torque and power.

Two sets of MRB design was considered by a varying working gap (MR-fluid) and keeping the outer disk radii and coil width constant. The torque value in MRB-I (having gap 1mm) was increased by 91.67% compare to MRB-II (having gap 2mm) at 0.6A. Experimental data were fitted to the theoretical equation of torque transmission. The torque value was increased as the MR-fluid working gap decreases.

The effect of variation in the outer disk radius and the coil configuration should be explored further in future studies according to the real-life applications.

Acknowledgment

The authors acknowledge Charotar University of science and technology for technical assistance and testing amenities.

References

[1] Wang J and Meng G 2001 Magnetorheological fluid devices: principles, characteristics and applications in mechanical engineering *P. I. Mech. Eng. L- J. Mat.* **215** pp 165-74

[2] Wereley N 2013 Magnetorheology: advances and applications *R. Soc. Chem.* London **6**

[3] Patel D M and Upadhyay R V 2018 Predicting the thermal sensitivity of MR damper performance based on thermo-rheological properties *Mater. Res. Express* **6** 015707

[4] Ahamed R, Choi S B and Ferdaus M M 2018 A state of art on magneto-rheological materials and their potential applications *J. Intel. Mat. Syst. Str.* **29** pp 2051-95

[5] Patel D M and Upadhyay R V 2020 Testing and evaluation of linear shear mode magneto-rheological (MR) damper based on rheological properties of MR fluid *J. Test. Eval.* **48**

[6] Jacob R, 1951 Magnetic fluid torque and force transmitting device United States patent US 2,575,360

[7] Park E, Stoikov D, da Luz L and Suleman A 2006 A performance evaluation of an automotive magnetorheological brake design with a sliding mode controller *Mechatronics* **16** pp 405-16

[8] Sarkar C and Hirani H 2013 Theoretical and experimental studies on a magnetorheological brake operating under compression plus shear mode *Smart Mater. Struct.* **22** 115032

[9] Nguyen Q H, Nguyen N D and Choi S B 2015 Design and evaluation of a novel magneto-rheological brake with coils placed on the side housings *Smart Mater. Struct.* **24** 047001

[10] Sohn J W, Jeon J, Nguyen Q H and Choi S B 2015 Optimal design of disc-type magneto-rheological brake for mid-sized motorcycle: experimental evaluation *Smart Mater. Struct.* **24** 085009

[11] Meyer R and Krueger D 2001 Minitab guide to statistics. Prentice Hall PTR

[12] Meeker D 2010 Finite element method magnetics FEMM **4**

[13] Li W H and Du H 2003 Design and experimental evaluation of a magnetorheological brake *Int. J. Adv. Manuf. Tech.* **21** pp 508-15

[14] Lee D Y and Wereley N M 1999 Quasi-steady Herschel-Bulkley analysis of electroand magneto-rheological flow mode dampers *J. Intel. Mat. Syst. Str.* **10** pp 761-69

[15] Nguyen Q H and Choi S B 2010 Optimal design of an automotive magnetorheological brake considering geometric dimensions and zero-field friction heat *Smart Mater. Struct.* **19** 115024

[16] Patel S R, Patel D M and Upadhyay R V 2020 Performance enhancement of MR brake using flake-shaped iron-particle–based magnetorheological fluid *J. Test. Eval.* **48**

[17] Fang F F, Liu Y D, Choi H J and Seo Y 2011 Core–shell structured carbonyl iron microspheres prepared via dual-step functionality coatings and their magnetorheological response *ACS Appl. Mater. Inter.* **3** pp 3487-95

[18] Patel S R, Upadhyay R V and Patel D M 2020 Life-cycle evaluation of anisotropic particle-based magnetorheological fluid in MR brake performance *Braz. J. Phys.* pp 1-9

Design, fabrication and analysis of compaction die for powder processing

Avadesh K Sharma[1*], Rajeev Singh[1], Arun K Tiwari[2], Ajay K Sharma[2]

[1]*Associate Professor, Department of Mechanical Engineering, Rajkiya Engineering College, Mainpuri, UP, India*
[1]*Research Scholar, Department of Mechanical Engineering, Rajkiya Engineering College, Mainpuri, UP, India*
[2]*Associate Professor, Department of Mechanical Engineering, Institute of Engineering and Technology, Lucknow UP, India*
[2]*Assistance Professor, Department of Mechanical Engineering, Institute of Engineering and Technology, Lucknow UP, India*

E-mail: sharma.avadesh@gmail.com, rajeevsingh150783@gmail.com

Abstract: A compaction die has an essential role in the powder metallurgy process due to the controlling and ease of processing. The shape and properties of die can significantly affect the features of the final part as the several steps and assessments are involved before design and fabrication of compaction die. This work is an effort to design and manufacture a compaction die that can successfully use for uni-axial compaction of NiTi powder. The design and development of die includes design consideration, 2D drawing, 3D model and processes involved in the fabrication of die. The die design was analyzed by ANSYS 19.1 software in a simulated environment, and the fabricated die was tested experimentally by preparing the composite sample successfully at 1000MPa compaction pressure using universal testing machine. Further, the compaction behaviour, density, compressive strength and hardness of developed NiTi composite have also evaluated.

Keywords: Powder metallurgy, compaction die, density, compressive strength, hardness

1. Introduction

Powder Metallurgy (PM) is a rapidly growing technique used for the development of most alloys and composites. It consists mainly three steps, as mixing of powdered materials, compress them into the desired shape, and then heating the compressed material in order to bind the material [1]. The PM process is also used to make specially designed material or part due to ease of processing and control, which are difficult to fabricate by other processes such as melting and forming [2, 3]. This process can prevent or significantly reduce the need for metal removal processes, thereby significantly reduce the manufacturing cost [4]. Another advantage of the PM process is that the components can be manufactured without decomposing or dissolving the material by avoiding the changes in material properties in the solid-liquid phase [5]. Therefore, powder processes are more flexible compare to other manufacturing processes such as casting, extrusion, and forging due to controlled characteristics of products prepared using this technique [6]. Several studies reported that the porous solids, aggregates and inter-metallic composites are efficiently manufactured by PM process with improved material characteristics [7]. There are several methods available for pressing in which uni-axial compaction is the simplest method of powder processing. In uni-axial compaction the powder is compacted by compress the metal powder in a close die chamber by applying high pressures in an axial direction [8]. Generally, the pressure of 150 MPa to 1000 MPa is usually used for metal powder compaction [9]. Typically the die is held in the vertical direction with the punch. The powder is then pressed into the desired shape and then ejected from the die chamber, followed by the sintering process at a specific temperature under controlled environment [10]. The significant limitations of pressure compaction process are higher porosity and circumferential micro-cracks developed on the surface of green compact due to friction generated between die wall and composite particles. In PM processed components, the micro-cracks are primarily originated during the compaction process and broadened during sintering [11-13]. Hence, present research work aims to design and develop pressure compaction die to compact NiTi composite powder with minimum porosity and micro-cracks. The design of die has analyzed by ANSYS 19.1 software under simulated conditions. Further, the performance of the die has been examined experimentally based on compaction behaviour of NiTi composite powder, porosity induced during compaction and mechanical properties by applying maximum pressure through a universal testing machine.

2. Material and methods

2.1 Die material

High carbon (C) and high chromium (Cr) cold work steel, generally known as D3 die steel was used for fabrication of compaction die. The studies shows that the higher hardness and wear resistance of die steel make it suitable for tooling applications such as powder metallurgy tooling, draw dies, and blanking and forming tools[13]. The compositional details of D3 steel is given in table 1. Singh et al. [14] have reported that the density and ultimate compressive stress of D3 steel is equal to 7.7 g/cm^3 and 2151 MPa. The hardness, tensile strength and yield strength of D3 steel is varies from 30 to 69 HRC, 990 to 2900 MPa and 705 to 2400 MPa respectively based on the condition of heat treatment [15].

Table 1: Chemical composition of D3 steel [13]

Element	C	Mn	Si	Cr	Ni	W	V	P	S	Cu
Content (%)	2.1	0.6	0.6	12	0.3	1	1	0.03	0.03	0.25

2.2 Design calculation

The compaction die was designed to develop the green compaction samples of NiTi composite with sample diameter 20 mm and thickness 5 mm. The design of compaction die is as follows:

Yield strength, σ_Y = 1300 MPa [15]

Chamber diameter, D_i = 20 mm

Outer diameter of die, D_o = 90 mm

Working load F = 320 KN

Cross section area $A = (\pi/4)*d_i^2 = 314$ mm^2

IOP Conf. Series: Materials Science and Engineering **992** (2020) 012005 doi:10.1088/1757-899X/992/1/012005

Stress developed in plunger due to applied load of 320 KN

$\sigma_1 = F/A = 320000/314 = 1019.1$ MPa [Hence ($\sigma_1 < \sigma_y$) safe working condition]

Stress developed in die due to load applied on plunger

$\sigma_2 = \sigma_1 [\{(R_o)^2 + (R_i)^2\}/ \{(R_o)^2 - (15)^2\}] = 1019.1[\{(45)^2 + (10)^2\}/ \{(45\}^2 - (10)^2\}]$

$= 1124.98$ MPa or 1125 MPa [Hence ($\sigma_2 < \sigma_y$) safe working condition]

2.3 Drawing of die

A 2D and 3D model of the die and punch set with the above mentioned design parameters was prepared using SOLIDWORKS modeling software. The 2D drawing of die chamber, upper and lower punch with dimensions is shown in figure 1(a). The 3D model of die assembly is shown in figure 1(b) and separate die parts are shown in figure 1(c). A suitable clearance of 0.5 mm has also given in the punch to reduce the surface contact between punch and wall of die cavity and thus minimizing friction and wear between the mating surfaces of punch and die chamber.

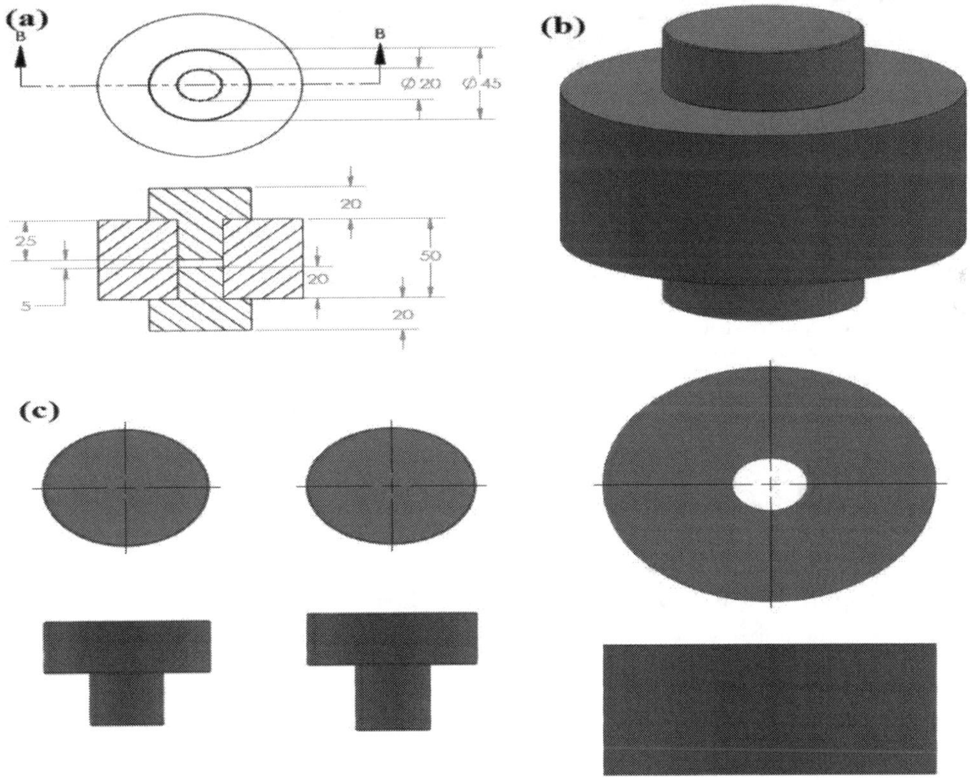

Figure 1: Schematic drawing with (a) 2D and (b and c) 3Dmodel of compaction die

2.4 Fabrication of die parts

A D3 die steel stock of 95 mm diameter and 50 mm diameter was procured from P. R store, Agra, India and cut into the required size for die chamber, upper and lower punch with power hack saw. The pre- machining of die chamber, upper and lower punch was done by facing, turning, drilling and boring operation on CNC lathe machine. Heat treatment of die cylinder and punch set was done in muffle furnace after the pre-machining in three steps as stress reliving, hardening and tempering. Stress relieving was done at 650 °C for one hour and cooled slowly. For hardening, the die parts were directly placed in a furnace preheated to 1000 °C and soaked for 25-30 minutes and then quenched in still air to harden it. The parts were placed in the furnace for tempering and the temperature of furnace is increased slowly to the desired tempering temperature (200 ^{0}C) and cooled slowly in furnace. Finally, the surface finishing operation of die parts was done using cylindrical grinder. The finished die chamber, upper and lower punch is shown in the figure 2.

Figure 2: Finished die parts (a) die chamber, upper and lower punch, and (b) assembly of die parts

3. Experimental analysis

3.1 ANSYS analysis

The designed die, upper punch and lower punch were analyzed for stress distribution using ANSYS 19.1. The analysis was selected as structural with the element type as round steel. The die chamber, both punches and NiTi powder were explored as .igs file from SOLIDWORKS 2017-18 and imported to ANSYS 19.1. The material properties of D3 steel and NiTi powder are given in table 2. All parts of die are meshed with each other and it is assumed that the 20 gram of NiTi powder was poured in die chamber. The pressure of 1000 MPa was applied on the punch and it displaced in downward direction inside the die chamber. The stress distribution of various die parts and compacted powder are shown in figure 3(a) to (e). The analysis shows that the upper punch and die chamber induced minimum stress equal to $1.9751*10^7$ Pa and the maximum stress ($1.1101*10^{10}$ Pa) was induced in lower punch and powder.

IOP Conf. Series: Materials Science and Engineering **992** (2020) 012005 doi:10.1088/1757-899X/992/1/012005

Table 2: Mechanical properties of D3 die steel and NiTi powder

Material	Density (g/cm^3)	Young's Modulus (GPa)	Poisson's Ratio
D3 steel	7.7	190	0.27
NiTi powder	6.75	75	0.33

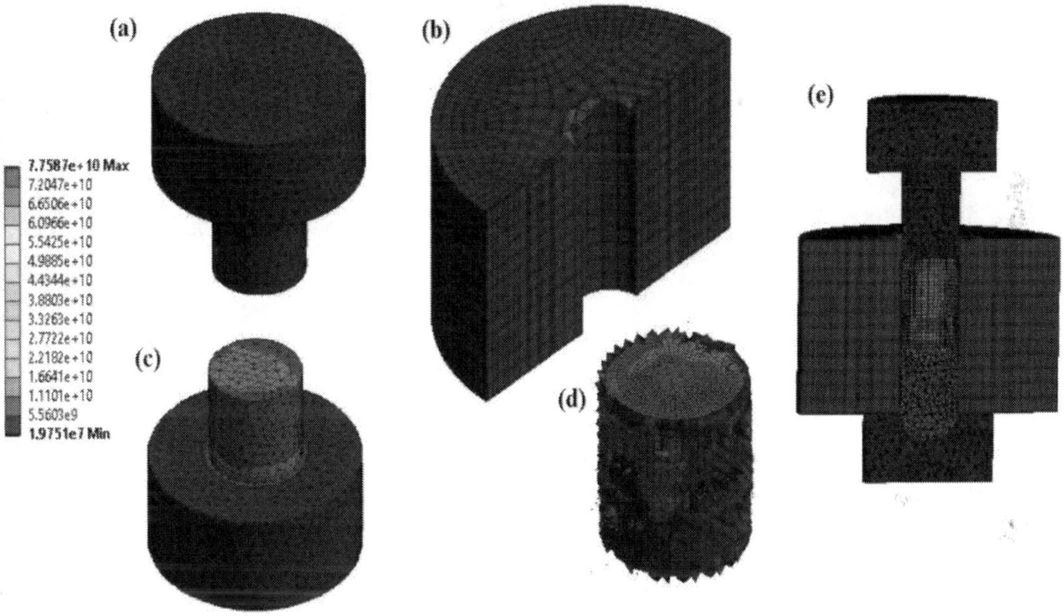

Figure 3: Stress distributions in (Pa) on (a) upper punch, (b) die chamber, (c) lower punch, (d) powder and (e) assembly

3.2 Powder compaction

The fabricated die was analyzed experimentally by compacting the NiTi composite powder using 400 KN universal testing machine in material testing lab at Rajkiya Engineering College, Mainpuri, India. After sufficient lubrication the 20 gm NiTi powder was poured in die cavity the powder was compressed by applying pressure of 1000 MPa at uniform rate. After the compaction, the compressed green compact was ejected from the die cavity with the help of ejector and green sample was sintered 1100 ^{0}C in muffle furnace for 4 hrs 30 min at the heating rate of 15 ^{0}C/min. The die arrangement in UTM and green sample developed after compaction and sintering are shown in figure 4.

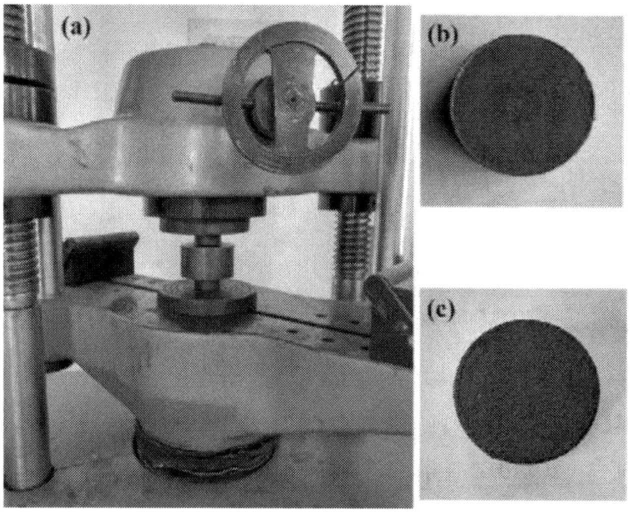

Figure 4: Image of (a) die arrangement on universal testing machine during compaction process and NiTi composite sample after (b) compaction and (c) sintering

3.3 Density

The green density of the sample was examined after compaction as well as sintering and compared with the theoretical density of NiTi composite powder. Theoretical density of NiTi composite powder was calculated using mixing rule. In addition, the green density of developed sample was measured using Archimedes principle in distilled water. The mass of the sample was measured on the electronic balance. The sample was attached to a thread and fully immersed in distilled water. The weight of the sample in distilled water was measured, whereby the loss of weight of the sample when suspended in distilled water is equal to the mass of fluid displaced, from which its volume can be calculated. Further, the porosity of sintered samples was analyzed by following formula:

$$\text{Porosity} = \{1 - (\text{sample density}/\text{theoretical density})\} * 100 \qquad \text{---- (1)}$$

3.4 Compressive strength and hardness

The compressive strength and hardness of sintered sample was also examined. Compression test has carried out using 400 KN universal testing machine at room temperature. The compressive strength NiTi sample is investigated by single experiment mode in which a cylindrical shape sample with 19.48 mm diameter and 12.8 mm thickness is used for the compression test. The set up for compression testing was shown in figure 5(a). To determine the hardness, the composite sample was tested using a Rockwell hardness testing machine with a diamond indenter. The Rockwell hardness of NiTi composite is measured at a load of 150 kg for 15 s dwell time. An average value of 5 measurements on the composite sample was determined. The set up for hardness testing is shown in figure 5(b).

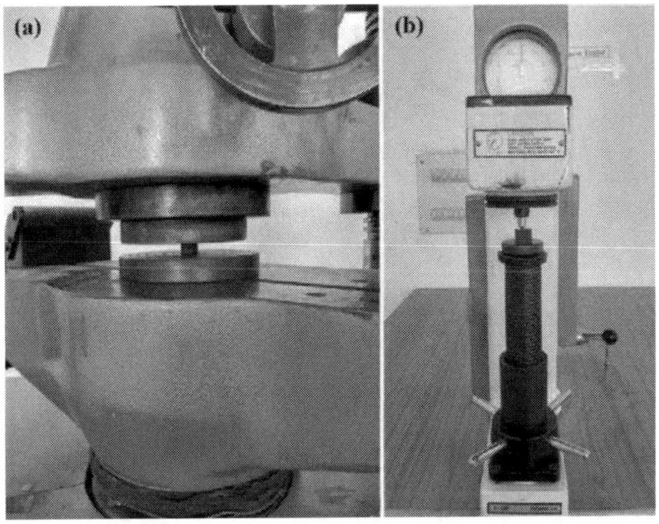

Figure 5: Arrangement for (a) compression testing and (b) hardness testing

4. Result and discussion

4.1 Compaction behavior

The compaction behavior of NiTi powder was directly influenced by compaction pressure. The compaction behavior of NiTi composite powder is shown in figure 6. It shows higher displacement of punch in the beginning of compaction due to the larger space available between the particles of composite powder but at the higher load displacement become constant at the end of compaction due to green strength and less space developed between composite particles. It is observed that the NiTi composite powder has successfully compressed to maximum density without any considerable die deformation.

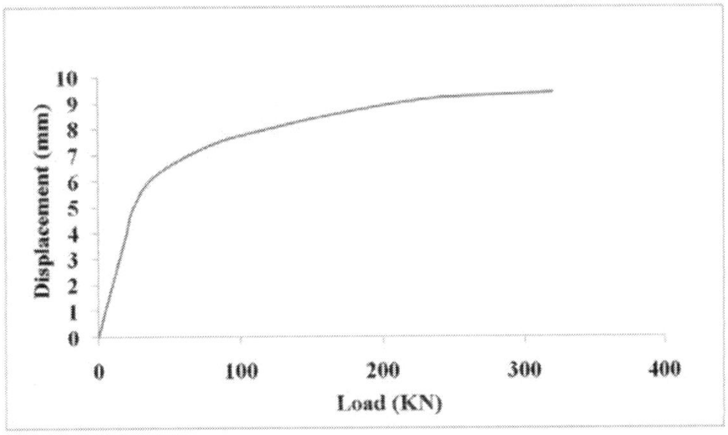

Figure 6: Compaction behavior of NiTi powder with respect to displacement versus load

4.2 Densification behavior

The density of developed sample was measured after compaction and sintering. The density of NiTi composite sample after compaction and sintering compared with theoretical density of NiTi composite is shown in figure 7. The results show that the density achieved after the compaction of sample (4.89 g/cm^3) is lower than the theoretical density (6.75 g/cm^3) of NiTi powder. The density of sintered sample (5.02 g/cm^3) was slightly increased due to the formation of necking between the NiTi particles and vaporization of moisture and lubricant. The porosity of sintered NiTi composite sample has calculated with the help of equation 1. By calculation, 27.56 % porosity was found during the compaction process.

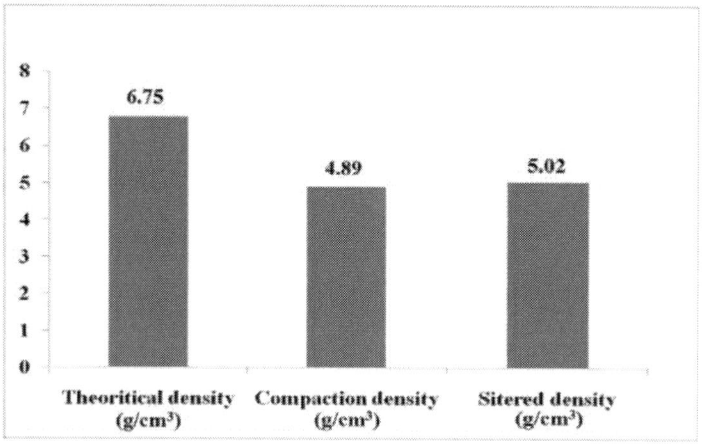

Figure 7: Density of NiTi powder and sample after compaction and sintering

4.3 Compression strength and hardness

Compression testing was performed on computerized UTM of capacity 400 KN. The sample was placed in between two flat plates and maximum failure load was recorded. The hardness test was performed on Rockwell hardness tester with diamond indenter. The result of compression and hardness test was shown in table 3. The results show that the sample prepared by compaction process by applying 1000 MPa load has achieved compressive yield strength, ultimate compressive strength and Rockwell hardness equal to 37.737 MPa, 115 MPa and 67 HRC respectively. It is observed that the compressive yield strength and ultimate compressive strength of NiTi composite is lower side due to induced porosity during the compaction process. However, the hardness of developed NiTi composite is near about of theoretical Rockwell hardness of NiTi alloy (58-65 HRC) [16].

Table 3: Compressive Yield strength, ultimate compressive strength and hardness of NiTi composite sample

	Yield stress (MPa)	Compressive strength (MPa)	Hardness (HRC)
NiTi composite	37.737	115	67

IOP Conf. Series: Materials Science and Engineering **992** (2020) 012005 doi:10.1088/1757-899X/992/1/012005

5. Conclusion

The compaction die was successfully designed and fabricated to develop NiTi composite by the application of 1000 MPa load on UTM. The ANSYS analysis shows the minimum stresses are distributed on die and upper punch and maximum stresses are distributed on powder and lower punch. The experimental analysis shows that the NiTi powder was successfully compressed to make green compact with 4.89 g/cm^3 sample density and 27.56% porosity. The sintering further improves the sample density due to formation of necking between the composite grains and vaporization of moisture and lubricant. The mechanical properties of sintered sample have optimum value compare with the theoretical values of compressive yield strength, ultimate compressive strength and hardness. Finally, it can conclude that the NiTi composite can successfully developed by applying uniform pressure upto 1000 MPa using cold compaction die.

Acknowledgement

The authors are grateful and acknowledged to Dr. A P J Abdul Kalam Technical University, Lucknow, India (Ref. No: AKTU/Dean-PGSR/2019/1570, Project Id: 286, Dated: 10/07/2019) for the financial and technical support under Collaborative Research and Innovation Program (TEQIP III) for this research work.

References

[1] Suryanarayana C and Al-Aqeeli N 2013 Mechanically alloyed nanocomposites, *Progress in Materials Science* **58(4)** 383–502
[2] Guo W, Liu B, Liu Y, Li T, Fu, A, Fang Q and Nie Y 2019 Microstructures and mechanical properties of ductile Nb-Ta-Ti-V refractory high entropy alloy prepared by powder metallurgy, *Journal of Alloys and Compounds* **776** 428–436
[3] Qian M, Yang YF, Yan M and Luo SD 2012 Design of low cost high performance powder metallurgy titanium alloys: Some basic considerations, *Key Engineering Materials* **520** 24–29
[4] Robertson IM and Schaffer GB 2010 Review of densification of titanium based powder systems in press and sinter processing, *Powder Metallurgy* **53**(II) 146–162
[5] Bolzoni L and Yang F 2019 Development of Cu-bearing powder metallurgy Ti alloys for biomedical applications, *Journal of the Mechanical Behavior of Biomedical Materials* **97** 41–48
[6] Pokorska I 2008 Deformation of powder metallurgy materials in cold and hot forming, *Journal of Materials Processing Technology* **196(1-3)** 15–32
[7] Kulkarni H and Dabhade VV 2019 Green machining of powder-metallurgy-steels (PMS): An overview, *Journal of Manufacturing Processes* **44** 1–18
[8] Parilak L, Dudrova E, Bidulsky R and Kabatova M 2017 Derivation, testing and application of a practical compaction equation for cold die-compacted metal powders, *Powder Technology* **322** 447–460
[9] Cristofolini I, Molinari A, Pederzini G and Rambelli A 2016 Study of the uniaxial cold compaction of AISI 316L stainless steel powders through single action tests, *Powder Technology* **295** 284–295
[10] Liu X, Hu L, and Wang E 2013 Cold compaction behavior of nano-structured Nd–Fe–B alloy powders prepared by different processes, *Journal of Alloys and Compounds* **551** 682–687
[11] Arifin A, Sulong AB, Muhamad N, Syarif J and Ramli MI 2014 Material processing of hydroxyapatite and titanium alloy (HA/Ti) composite as implant materials using powder metallurgy: A review, *Materials & Design* **55** 165–175
12] Kang CS, Lee SC, Kim KT and Rozenberg O 2007 Densification behavior of iron powder during cold stepped compaction, *Materials Science and Engineering: A* **452** 359–366

[13] Lee SC and Kim KT 2008 Densification behavior of nanocrystalline Titania powder under cold compaction, *Powder Technology* **186(I)** 99–106
[14] Singh RB, Seema and Kumar D 2019 Heat treatment of tool steel D3 and effects on mechanical properties, *International Journal of Recent Scientific Research* **10(V)** 32540–32545
[15] Ghazi SS and Mashloosh KM 2015 Influence of heat treatment on resistance of wear and mechanical properties of die steel kind D3, *American Journal Of Scientific and Industrial Research* **5(II)** 33–40
[16] Chen H, Zheng LJ, Zhang FX and Zhang H 2017 Thermal stability and hardening behavior in superelastic Ni-rich Nitinol alloys with Al addition, *Materials Science and Engineering: A* **708** 514–522

IOP Conf. Series: Materials Science and Engineering **992** (2020) 012006 doi:10.1088/1757-899X/992/1/012006

Onset of detonation through a cold flow analysis using metallic diaphragms – an experimental approach

S Srikrishnan[1], T Shaafi[2] and P K Dash[3]

[1]Department of Mechanical Engineering, Institute of Aeronautical Engineering, Hyderabad.
[2]Department of Mechanical Engineering, Saveetha School of Engineering, Saveetha Institute of Medical and Technical Sciences, Chennai. India.
[3]Department of Aeronautical Engineering, NITTE Meenakshi Institute of Technology, Bangalore.

Abstract. Modern era is looking for high speed transportation systems across globe with due comfort. Innovative design in both aerodynamic and propulsion studies have been made for achieving supersonic to hypersonic speeds ranging a claim of 1km/s under stratosphere limit. The complexity in designing such vehicles in a high-altitude test facility is borne due to the fact that simulation of such conditions as low density, no shock conditions, high rate of dissipation and shock flow comes from unavailability of sufficient technology to pressurize gases to the optimal testing conditions. The prominent ways of constructing a shock tube either involve a blow down type or a vacuum suction type. Here a blow down type shock tube require a built-up pressure comparable to the velocity at the test section. Fibre diaphragms blow out at less critical pressure and are unsuitable for high speed flows. Metal foil are incepted as a concept of diaphragms for this particular problem. Aluminium foils are stacked across a mounting flange on a settling chamber to determine the maximum pressure at which blow down occurs. The results are tabulated for an incremental pressure schema of 1 bar to 10 bar with which it was found that the wave pressure development occurs during each batch of aluminium and recorded to be between 3.5 bar and 9 bar the successive velocities at the exit was found to be increasing with increase of blast pressure.

1. Introduction

Characterization of shock train inside a shock tube by varying the shock tube length, diameter of the diaphragm and pressure ratios has been extensively studied. Investigations were conducted to visualize the interaction between the reflected shockwave and the contact surfaces. The resulting wave expansion produces a supersonic wave which is then decelerated and interacted to produce stronger shockwave [1]. Studies on numerical simulations of flow characteristics in micro shock tube by observing shockwave surface propagation and attenuation has been done at different condition resulting from viscous effects, the friction between shockwave and tube wall aiding in forming boundary layer and micro shocks. Increment of diaphragm pressure ratios have indicated an increment of the Mach number at the exit as well as the shockwave strength. So it can be concluded that higher pressures are required at the settling chamber which affect the exit velocity through jet flow control equation given in equation 1 [2].

$$\frac{dV}{V} = -\frac{dA}{A}\frac{1}{1-Ma^2} \tag{1}$$

Further, shockwave interaction with boundary walls due to the high pressure and velocity involve heat transfer from shock to the wall of pipe. Experimental investigation using various metal foils of significant thicknesses were used to test the wear rate due to the thermal and fluid boundary layer interaction between the shockwave and surface of the expansion tube. It has been noticed that the heat transfer through the interference or blockage is higher at high Reynold's number and the erosion effect on the plates were experimentally measured to have a significant change in the demography of the blockages [3]. Fabrics and other materials are also considered for choking the flow from the settling chamber to the jet pipe. At high pressure, singular layers of fabric usage become irrelevant as the cloth permeability ratio is higher and the conversion of subsonic to supersonic flow is generally not possible. However usage of multiple layers of cloth with different orientations make it almost feasible similar to that of metal foils and tracing paper which have been better diaphragm designs under high pressure conditions and has low permeability ratios [4].

Shock flow conditions have also been studied by researchers to describe the flow evolution in an experimental investigation along a flat plate. It has been found that flow structure is important factor which determine the flow evolution and a reflected shockwave may affect the total pressure gradient inside the jet flow thereby inhibiting the flow characteristics stating that heating of gases flowing through the jet occurs due to interaction between fluid, thermal boundary layers and reflected shockwaves [5]. Various inert and standard gases have also been studied inside shock tubes for their sustainability at high pressures due to the compressibility factor of such gases. The gases shows that shockwave formation goes through multiple stages and the pressure gradient always does not follow an incremental schema. The high pressure ratio occurs after the first rise of shockwave due to the propagation effect and the interaction between compressed gases and air. Auto ignition of hydrogen with ambient air at high pressure has been utilized to visualize the effect of compressed gases and the minimum pressure for the auto ignition and subsequent pressure wave has been studied to give an idea about minimum pressure at which the diaphragm should rupture for the maximum exit flow velocities to be achieved and further the type of material must be restricted to low permeability ratio materials for achieving maximum efficiency of the shock tubes [6,7].

Acoustic relays play a significant role on measuring the pressure wave strength of shockwaves occurring from diaphragm ruptures as flow velocity measurement instrumentations become difficult to construct based on pressure based load cell as the probes for collection of pressure data are themselves considered be micro blockages or cavities in the flow stream as it is affecting the demography of the flow by creating small perturbations in flow stream [8–10]. The small perturbations causes a significant effect in the downstream boundary flow interaction between shockwave and pipe wall where flow mitigation becomes imminent and slipstream reduction occurrence possibility increases. PCB based microphones have been extensively used at multiple lengths and angles across the experimental flow stream to record the soundwave emission and such recordings are then converted to pressure wave strength using a mathematical formulation[11-13].

Mylar 30 woven fabric were also tested as a possible materials for diaphragms in shock tubes and have been found to be resistant to micro cracking for nominal pressures[14]. The panel responses and growth rate of crack have been measured by high rated moving cameras to see the matrix crack fabrication both on and below the surface of the multiple layer diaphragms. Further, it is apparent that rigid diaphragms partitions are better than flexible and expandable materials as they prevent seepage of air pressure at higher pressures [15]. Foils have been used in this study to have the advantages of rigid diaphragms as sealing qualities of a flexible diaphragm material by a defined design change of the mounting flange by replacing normal flange by mating flange arrangement.

2. Design and Methodology

The experimental methodology involved in the testing on oblique shockwaves have a multitude of constraints. The major constraint being physical as the production of shockwaves involved generation

IOP Conf. Series: Materials Science and Engineering **992** (2020) 012006 doi:10.1088/1757-899X/992/1/012006

of high settling chamber pressure and maintaining it for the duration of gradual increment till the blast pressure is achieved without disturbing the structural integrity of the settling chamber. Further, high pressure expansion waves generated post blast from the settling chamber expands throughout the delivery pipe or test section in a blowdown shock tube which travel at high speeds causing higher acoustic disruptions in the surrounding areas and presence of any debris in the delivery tube itself will cause such objects at high speeds causing severe damage to property and life.

It was ensured that proper care has been taken during the entire experimentation process so as to avoid any damage either to property or life during the entire duration of the experimentation. Also, the experiments conducted were in a closely monitored conditions wherein, hot flow analysis were avoided to prevent the dangers of expanding combustion gases in closed vicinity and the cold flow can be calibrated to the hot flow analysis using a chemical kinetic model to ensure continuity and repeatability.

The experimental setup has been ensured to withstand high pressures at the settling chamber and to manage the range of pressure inside the chamber, cylindrical structure has been used for even distribution throughout the volume. The thickness of the volume is maintained around 12mm of solid steel in all directions to prevent gaseous expansion even through micro cracks and holes in the surface of settling chamber. Proper welding methodologies are implemented to ensure blockage of micro holes in the surface where the pressure gauge, inlet flange and delivery flanges are attached to ensure maximum pressure building capacity of the settling chamber. A mating flange setup is given to ensure that the leakage of air is not possible through the diaphragm hole to atmosphere and a microphone is attached near the flange plate to ensure radio silence during the surging and pressurizing of the settling chamber before the onset of the blast.

Diaphragm design for micro shock tubes were particularly difficult reason being the economy involved behind the material selection and availability. Soft material selection is generally less that to in metallic components. Foiled diaphragms are an option but metallic foils are also limited to availability and manufacturing. Food grade aluminium foils aids in becoming the best alternate for micro shock tube applications. Further a micro experimental shock tube can also become a detonation tube with the inclusion of high energy materials (fuels).

The onset of such complex problems at low budget is defined in this experimental process where in stacks of compressed aluminium foils are being moulded to a exit flange of settling chamber. Here the settling chamber is pressurized by an external reciprocating air compressor and the air expanding blast open the diaphragm to evolve super-sonic jet flow wave propagation and velocity assimilation due to wave propagation are detailly studied by experimental and numerical simulations in a standard length tube with standard tube diameter aiding in the validation of pressure expansion phenomena in a tubular flow.

The layers of the aluminium foils are then compressed in a standard UTM machine where the air gap between the layers are diminished or completely worn out so that the wave expansion modular will work perfectly. The stacks of aluminium are then cut for the exact shape of the flange and placed in between the mating parts. Experimentation is done in lieu with regression curve based on the data assimilated from the initial analysis between number of layers and explosion pressures. The results evolved aid in selection of number of layers with reference to the pressure of blasts. The instrumentation required for wave pressure modelling in an experimental method is both economically and technologically sound. Here a new instrumentation has been developed where wave pressure has been recorded through acoustic frequencies which can further be numerically endowed as velocity and jet flow. The formulation has been given in Equation 2.

$$L_P = 20log_{10}\left(\frac{p}{p_0}\right) dB \tag{2}$$

Where L_P is the frequency of sound in decibels which are found out using the acoustic frequency recording which comprises of attachment of probes on the shock tube at various angles such that to clearly analyse the wave propagation audibility from all the directions of the shock tube at various lengths such as 0mm, 25mm, 50mm, 75mm, 100mm at various angles such at 0°, 90°, 180°, 270°, 360° and P is the outcome pressure of the wave propagation at the above positions and P_o is the reference pressure that is considered as the sound pressure that is typically the threshold of human hearing that is $2*10^{-5}$Pa. The experimental setup to determine the onset of detonation is shown in the figure 1.

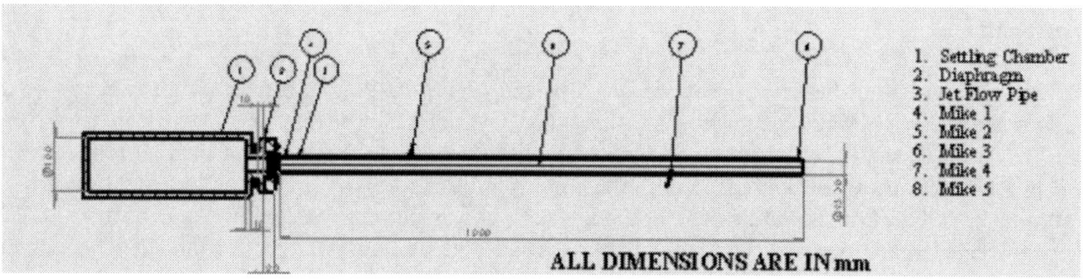

Figure 1. Experimental Setup

3. Result and Discussion

The experimental investigation of the detonation initiation by cold flow across blowdown from a diaphragm isolated settling chamber and a discharge tube yields an acute understanding of the behaviour of gases at high compressed state and during rapid discharge. The acoustic redundancy was maintained by a noise cancellation test prior to the actual experimentation, wherein an initial test through the pipe was used to record the noise normally occurring at the free state and such frequencies were then removed from the recorded frequencies to maintain the calibrated frequency and then the wave pressure were calculated with reference to the aforementioned equation. It is noted that during the expansion of the pressure wave across the length of the delivery tube there is a continuous decrement of wave pressure signifying expansion of gases through the delivery pipe with rapid increment to attain maximum exit velocity ranging in a supersonic delivery flow. The relationship between the burst pressure and the exit wave pressure are shown in Table. 1.

Table 1. Relation between Acoustic Readings and Wave Pressure.

S. No.	Burst Pressure (bar)	Acoustic Readings (dB)					Wave Pressure (bar)				
		Mic 1	Mic 2	Mic 3	Mic 4	Mic 5	P1	P2	P3	P4	P5
1	1.213	195.6	195.3	195.4	195.2	195.1	1.2	1.16	1.17	1.15	1.13
2	2.913	203.2	203	202.7	202.8	202.6	2.89	2.82	2.72	2.76	2.69
3	4.513	207.06	206.8	206.4	205.9	205.6	4.5	4.37	4.17	3.94	3.81
4	5.913	209.4	209.1	208.4	208.1	207.4	5.97	5.7	5.26	5.08	4.68
5	7.213	211.1	210.9	210.8	210.6	209.7	7.17	7.01	6.93	6.77	6.1
6	8.713	212.7	212.5	212.2	212.3	209.9	8.63	8.43	8.14	8.24	6.25
7	10.013	213.9	212.8	211.6	210.8	210.4	9.9	8.73	7.6	6.93	6.666

Using the above table the exit velocity of such blasts where the expansion of gases leads to higher exit Mach number can be calculated with reference to the pressure wave velocity equation. It is noticed that the settling chamber burst pressure is directly proportional to the exit velocity during the supersonic exit phenomena and a stalling is occurring at the higher blast pressure due the back pressure effect due to very low pressure at the exit of the delivery pipe. Since the onset is a cold flow analysis, ionization flow and thereby hypersonic velocities were not recorded and cannot be expected at such low onset pressures. The minimum blast pressure required for a hypersonic flow at the settling chamber may be predicted using the incremental schema of the above table and may be around 16 bar in a cold flow analysis and range around 14 bar using a hot flow condition. The delivery pipe is ensured to have no contaminants or debris to ensure the pressure expansion is even. A sample of the ruptured diaphragm is shown in the Figure 2. below and it shows that the diaphragm rupture is matching with the delivery tube and no material loss does not occur suggesting that the multilayer aluminium foil diaphragm has better structural integrity when compared to composite layered diaphragms and are better suited for hot flow detonation initiation tests.

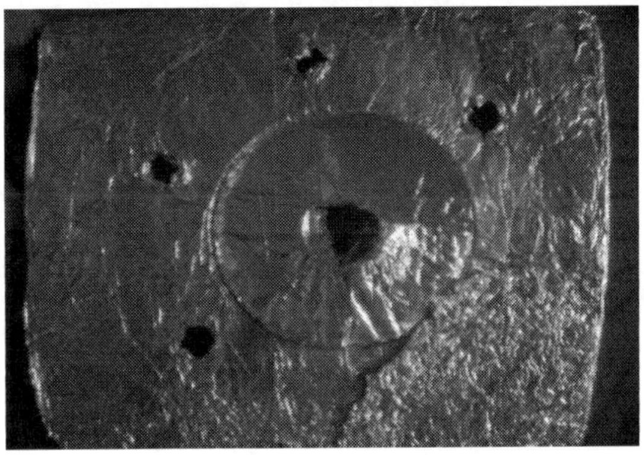

Figure 2. Ruptured Diaphragm

Using the table we may calculate the various exit velocity values and significant Mach numbers at which the flow is expanding from the isentropic expansion flow equation. The comparative graphs for exit velocity and burst pressure is shown in Figure 3.

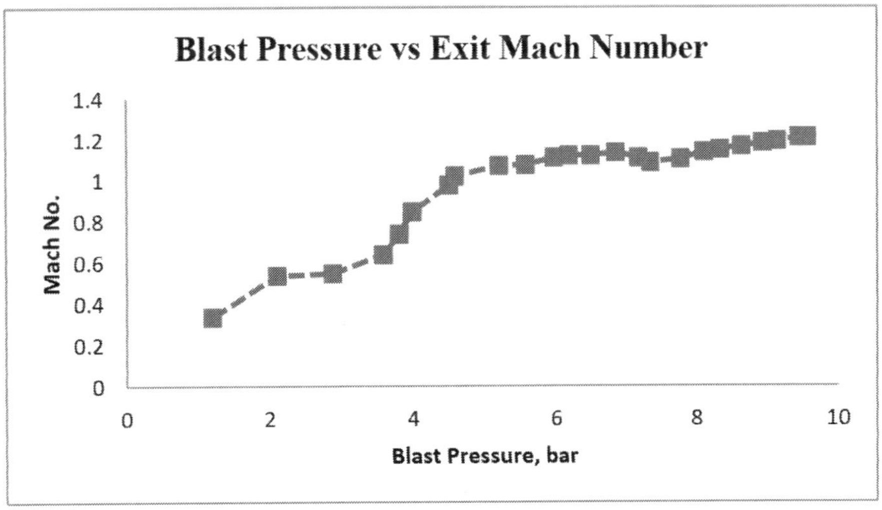

Figure 3. Relationship between Blast Pressure and Exit Mach number.

The relation between blast pressure and exit Mach number signify that the maximum velocity at the exit is around 1.4 Mach or 476 m/s in a cold flow isentropic expansion flow conditions where the working fluid is compressed air for a burst pressure of around 10 bar and the corresponding number of layers of aluminium foil required to sustain such high blast pressures in the settling chamber is around 26. The equivalence to exit pressure to the number of layers of aluminium foil required to sustain the pressure is shown in Figure 4 and may be utilized to predict the number of layers of aluminium foil required during a combustion to detonation onset analysis.

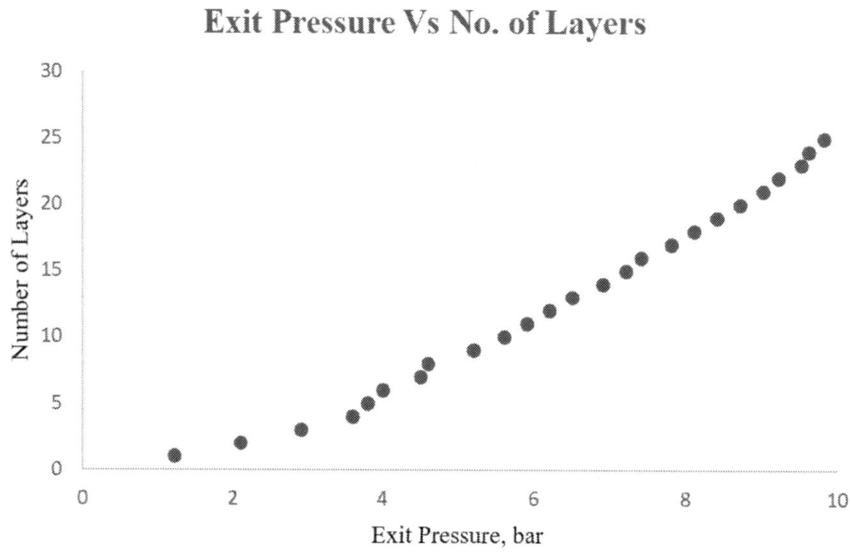

Figure 4. Exit pressure vs number of layers of aluminium foil.

4. Conclusion

With reference to the experimental investigation of onset of detonation using a cold flow analysis through multiple layers of metal diaphragm we can conclude the following important points.

1. Food grade aluminium foils stacked in multiple layers and bonded together at high physical loads can be used as effective diaphragms in both cold and hot flow conditions.

2. A predictive model can be realised using the presented data for calculating the thickness of the diaphragm and number of layers required for conducting hot flow experiments.

3. It is found that a maximum velocity of 476 m/s or 1.4 Mach number of exit pressure is possible for a settling chamber diaphragm rupture pressure of around 10 bar.

4. During the expansion of gases through the delivery tube there is a stark decrement phenomena of exit pressure signifying expansion of gases in the absence of any blockages in the delivery tube and an incremental velocity schema.

IOP Conf. Series: Materials Science and Engineering **992** (2020) 012006 doi:10.1088/1757-899X/992/1/012006

5. References

[1] Dong Wook Kim, Tae Ho Kim, 2017 *A study on characteristics on shock train inside a shock tube, Journal of Theoretical and Applied Mechanics* **7** pp. 366-371.

[2] Guang Zhang, Toshiaki Setoguchi and Heuy Dong Kim, 2015 *Numerical Simulation of Flow Characteristics in Micro Shock Tubes Journal of Thermal Science* **24** pp. 246-253.

[3] Serge M. Boldyrev , Volf Ya. Borovoya, Andrey Yu. Chinilov, Victor N. Gusev, Serge N. Krutiy , Irina V. Struminskaya , Larissa V. Yakovleva , Jean Délery , Bruno Chanetz, 2001 *A thorough experimental investigation of shock/shock interferences in high Mach number flows* Aerospace Science Technology **5** pp.167-178.

[4] Olivier Lamberts, Philippe Chatelain, Yann Bartosiewicz, 2018 *Numerical and experimental evidence of the Fabri-choking in a supersonic ejector, International Journal of Heat and Fluid Flow* **69** pp. 194–209.

[5] Alexey Kiverin, Ivan Yakovenko, 2018 "On the mechanism of flow evolution in shock-tube experiments", Physics Letter A, *https://doi.org/10.1016/j.physleta.2017.11.033*

[6] N. Kitabayashi, Y. Wada, T. Mogi, T. Saburi, A.K. Hayashi, 2013 *Experimental study on high pressure hydrogen jets coming out of tubes of 0.1-4.2 m in length* International Journal of Hydrogen energy **38** pp. 8100-8107.

[7] Qiangling Duan, Huahua Xiao, Wei Gao, Xiaobo Shen, Qingsong Wang, Jinhua Sun, 2015 *Experimental investigation on shock waves generated by pressurized gas release through a tube*, Journal of Loss Prevention in the Process Industries, **36** pp. 39-44.

[8] P. Laffay, S. Moreau, M.C. Jacob, J. Regnard, 2018 *Experimental study of the noise radiated by an air flow discharge through diaphragms and perforated plates*, Journal of Sound and Vibration, **434** pp. 144-165.

[9] M. Pankow, B. Justusson, A. Salvi, A.M. Waas, Chian-Fong Yen, Seth Ghiorse, 2011 *Shock response of 3D woven composites: An experimental investigation*, Journal of Composite Structures, **93** pp. 1337-1346.

[10] Haselbacher and S. Balachandar, 2007 *Open-Ended Shock Tube Flows: Influence of Pressure Ratio and Diaphragm Position* AIAA Journal, **45**.

[11] Hyoung Jin Lee, Yeong Ryeon Kim, Sei-Hwan Kim, In-Secuk Jeung, 2011 *Experimental investigation on the self-ignition of pressurized hydrogen released by the failure of a rupture disk through tubes*, Proceedings of Combustion Institute, **33** pp. 2351-2358.

[12] Shachar Berger, Oren Sadot, Gabi Ben-Dor, 2010 *Experimental investigation on the shock-wave load attenuation by geometrical means* Journal of Shock Waves, **20** pp. 29-40.

[13] J. Panda and R. G. Seasholtz, 2002 *Experimental investigation of density fluctuations in high-speed jets and correlation with generated noise* Journal of Fluid Mech., **450** pp. 97-130.

[14] V. Aune, E. Fagerholt, K.O. Hauge, M. Langseth, T. Børvik, 2015 *Experimental study on the response of thin aluminium and steel plates subjected to airblast loading* International Journal of Impact Engineering, *http://dx.doi.org/doi: 10.1016/j.ijimpeng.2015.11.017.*

[15] M. Stoffel, R. Schmidt, D. Weichert, 2001 *Shock Wave Loaded Plates* International Journal of Solids and Structures, **38** pp. 7659-7680.

Design of multistage forging processes by numerical simulation and experimental verification

P Ruban [1], J Jayaprakash[1] and S Surendarnath[2]

[1] Department of Mechanical Engineering, Saveetha School of Engineering, Saveetha Institute of Medical and Technical Sciences, Chennai, India.

[2] Department of Mechanical Engineering, Nalla Narasimha Reddy Education Society's Group of Institutions, Hyderabad, Telangana, India.

E.mail: agastinruban@gmail.com

Abstract : The Variety of components with different shapes & Sizes with different materials can be formed using Multi stage hot forging processes. In previous times, forging process in multi stage was designed on basis of experience gained in the previous forging practices. Nowadays, many software packages based on FE (finite elements) and FV (finite volumes) method are developed. This enables creation of forging process in more effective and accurate condition. The paper insight is given in the design of multi stage hot forging process for a critical geometry of Socket Forging in a multi-step forging process. The main criterion is to minimize the number of forging stages, proper filling and stressing of the dies by analysing the contact pressure. The Theoretical derivation & Numerical Simulation results are verified with practical, industrial environment.

Keywords: Hot forging process, Multi-stage Forging Process, DEFORM.

1. Introduction

Forging is a metal forming processes which is the key industrial production technologies that can create parts which are stronger than any other manufacturing processes. Components that are produced by forging have a high structural integrity, provide good fatigue life, maintain uniformity and withstand high impact strength. The reliability and human safety are also highly demanded. The High flexibility in the process provides the act of manufactured the great variety of different profiles (Size & Shapes) and materials like steel, aluminum, titanium, alloys.

Every year different forged components are produced in worldwide. In the most cases those components are assembled into automobiles, airplanes, vehicles, ships, missiles and others. The study case says that the modern car design includes 250 forged components. In average, large airplane contains over 18000 high precision forged components [1].

The Current field of forging technology is characterized by a high research work in a number of different areas such as metallurgy, material science, heat treatment, tooling design, environmental issues, etc. Number of papers has elaborated the problem of forging-set up [2, 3, 4, 5, 7]. Complex profiles or geometries are produced in multi-stages [8,9]. In such cases it is essential to determine optimal process design which has a minimum number of forging stages. For a long time set-up of process, Sequences has to be experimentally generated by trial and error method which is highly time and cost consuming. This also inevitably reflects high on forging quality and costs [6].

Instead to those problems, the greater solution is numerical simulation and optimization, which has been grown rapidly in the last years, This software packages based on FE (finite elements) and FV (finite volumes) offers a high possibility to perform forging multi-Stage process designing in more efficient and qualitative way.

In this paper Optimized design of Multi Stage hot forging processes of a complex profiled socket are presented in terms of minimal number of forging stages, complete die filling and minimum material waste. Verification of the obtained and optimized solution has been performed in production environment forging plant, using Friction Drop hammer.

2. Hot forging technology of socket forging

Socket Forging as shown in figure.1, made from Alloy Steel Material EN-8D (BS Standard)[14][15], is a Forged component which is used in the steering mechanism of a vehicle. Due to the high dynamic load getting exploited in the part is produced by forging and by additional Heat Treatment Process for the high load withstanding. The part is also included in the metal cutting operations for the assembly requirement.

The Material in extra is added with the billet for the proper filling of the Cavity which is provided in the Dies. This Excess material which is added with the billet is known as the Flash. A flash is used as a carrier for the impact to blow from the cavity of the die restricting the outward direction flow of the metal which helps in the filling of all such thin ribs and bosses in the upper die cavity. The Excess flash gets to flows along the forging parting line of the Socket Forging considered. The flash is to be provided uniform around the Socket forging peripherals around the parting plane. The minimum flash allowances suggested are given in equation 1. [8]

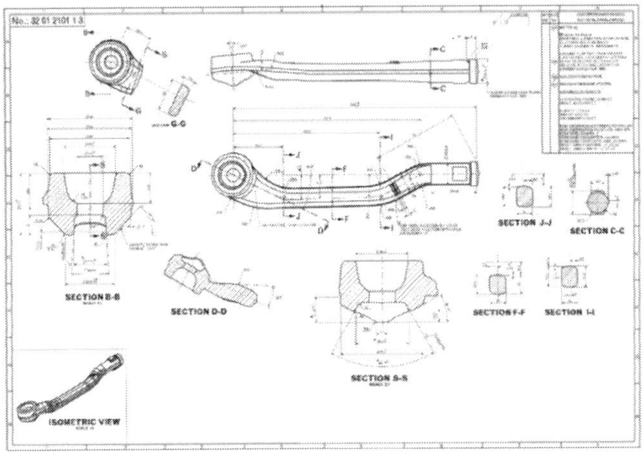

Figure 1: Socket forging drawing

Tf = 1.13 + 0.89 W 0.5 – 0.017 (W)...................... (1)
Wf/Tf = 3 + 1.2 e (–1.09 W)

Where Tf = is flash thickness in mm
Wf/Tf = is flash land ratio
W = is weight of forging in kg

There for thickness of flash for Socket Forging is:

Tf = 1.13 + 0.89 (0.6) 0.5 – (0.0170 X 0.6)
= **1.979 mm**

Flash land ratio

Wf/Tf = 3 + 1.2 e -1.09 X (0.6)
= 3.624

Therefore Wf = 3.624 X 1.979
Wf = 7.172 ~ 7.2 mm

With addition of flash provided, A Provision for the material to get flowed outward from the cavity should also be given and this helps in the proper closed stroke of the dies which is known as gutter. This gutter should always be provided more than the flash getting outward from the billet. The most preferred gutter size with the comparison of billet size is getting presented in the table 1. The flash land provided in the die should always be more than 3% of the maximum forging thickness of the Socket Forging [12]. If the flash land is provided less than that, then the Stroke energy required for the forging process get increased because of the excess flash getting trapped in the finishing impression of the die, Also the flash land gets wear out quickly and also the cavity remains unfilled.

Table 1: Flash and Gutter Table

Billet size(mm)	Flash		Gutter	
	Width (mm)	Thickness(mm)	Width (mm)	Thickness(mm)
Up to 35	4.5	0.8	25	3.0
36 to 50	5.3	1.0	25 to 32	4.5
51 to 65	6.5	1.5	32 to 38	4.5
66 to 75	8.0	2.0	32 to 38	4.5
76 to 100	10.0	3.0	38 to 44	6.5

The Maximum Width and Thickness of the Socket Forging is:

Tmax = 28.5 mm
Wmax = 245 mm

Therefore the area of Cross Section of the billet is:

= 6982.5 mm2

By adding the Flash Thickness and Width

= 28.5 + 1.979
= 30.479 mm

Width = 245 + 7.2
= 252.2 mm

The Total Area of Cross Section of Billet is:

= 30.479 X 252.2
= 7686.8 mm^2

Approx. Volume of Socket forging is:

$$V_c = \frac{\pi \times 40^2}{4} \times 28.5 + \frac{\pi \times 19.3^2}{4} \times 51.5 + (15.5 \times 19.1 \times 180)$$

V_c = 104169.634 mm 3

Volume of flash:

V_f = 1.979 x 7.2 x 245

= 3490.956 mm^3

Adding volume of forge and volume of flash we get

= 104169.634 + 3490.956

= 107660.590 mm3

By dividing the volume with area of billet to get the length

D = 107660.590 / 3490.956 = 30.84

D ~ 31 mm

So, it is a primary need to provide enough provision of about 10 to 15 mm between the both die impressions. If the space is provided less, upsetting of the die block will not take place properly. The clearance between the two die cavities depends on the depth of the die cavity and is given as follows:

Die Cavity depth (h) = 0.5 T

Where T is the Maximum thickness of Socket Forging

(h) = 0.5 x 28.5

= 14.25 mm

Cavity Spacing W = 3.1 X (h) 0.7

W1 = 3.1 X (14.25) 0.7

= 19.91 ~ 20mm

Die Cavity distance at edges is given as

(E) = 3.4 X (h) 0.76 = 3.4 X (14.25) 0.76

(E1) = 25.61 ~ 26 mm

The realization of the Socket forging is done by the designed dies as per the multi-stage principle, i.e. complete stages in one single die. Process of socket forging was simulated by DEFORM 3D software [7,10,11,13,16] and the forming of all stages of work piece. The initial data considered for the multistage process are as follows:

- Billet rod - 80mm in diameter and 160mm long
- Preheating temperature of the billet: 1250°C
- Die preheating temperature: 400°C
- Hammer capacity: 80KJ

2.1. Operation 1 – Free Upsetting:

This is a preparatory step in which billet (rod) is compressed on one side for the amount of 32mm as shown in figure.2a & 2b. This operation is foreseen for the mass distribution along the longitudinal axis of the work piece in order to enable better die filling in successive [11], following operation.

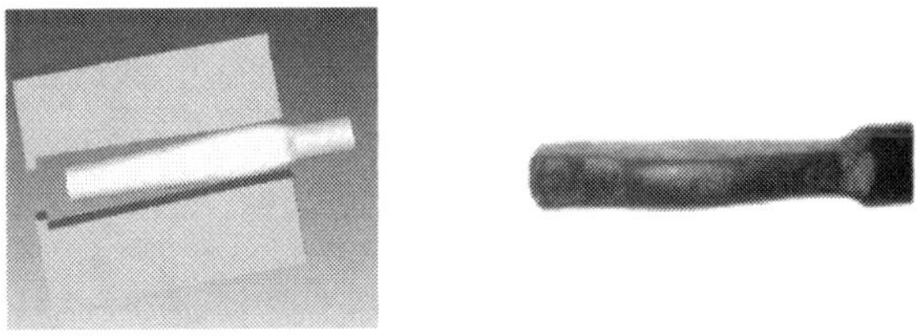

Figure 2: Free upsetting Forging a) CAD Deformation b) Actual Formation

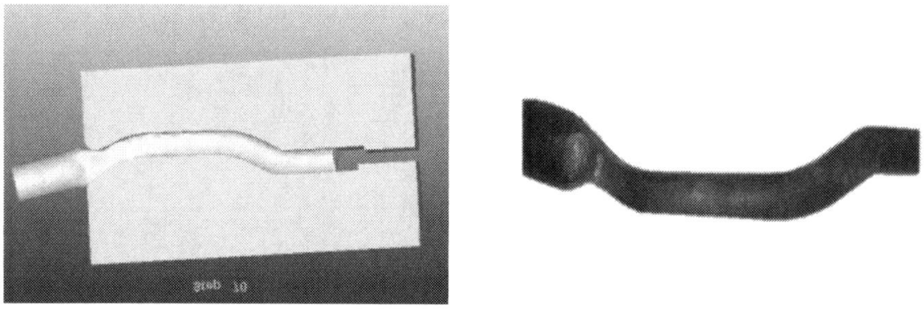

Figure 3: Bending Forging
a) Cad Deformation b) Actual Formation

2.2. Operation 2 – Bending Forging:

This operation is done to form the bending profile in socket forging as shown in figure 3a & 3b. The bended part is placed in the next stage in order to get the part filled. For the proper contact of the billet in the preforming stage, the bending operation is carried out. This process will also eliminate the more material usage.

2.3. Operation 3 – Preforming Forging:

This operation is designed in order to prepare the work piece for the final forging step, i.e. for the creation of preformed forging shape [10]. In this way in final die cavity only minimal filing of required profile of the work piece geometry is to be imposed.

Preforming operation is performed in special preforming die as shown in figure.4a & 4b. Operation is completed after 3 strokes of hammer. Die filling after this operation is 100%. Contact stresses at the work piece after preforming operation of cross-section (figure.4a & 4b) are significantly lower than the allowed values (maximum 786 MPa), which is a good starting point for final forging. Maximal allowed values of contact stresses in hot forging are 2000 MPa.

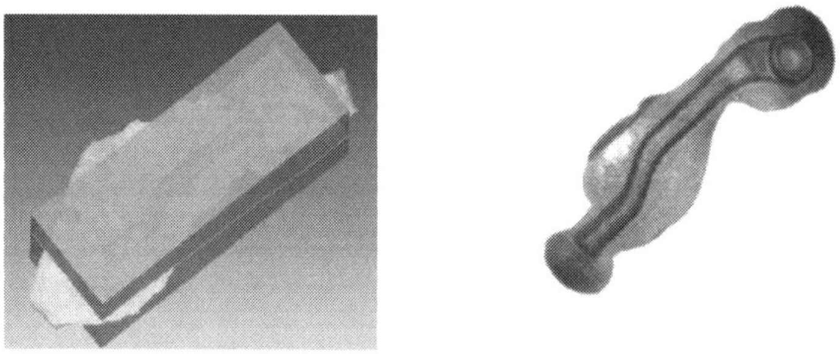

Figure 4: Preforming forging
a) Cad Deformation b) Actual Formation

IOP Conf. Series: Materials Science and Engineering **992** (2020) 012007 doi:10.1088/1757-899X/992/1/012007

2.4 *Operation 4 – Final Forging*

In final forging operation work piece obtains the final shape and dimensions as shown in figure.5a & 5b.

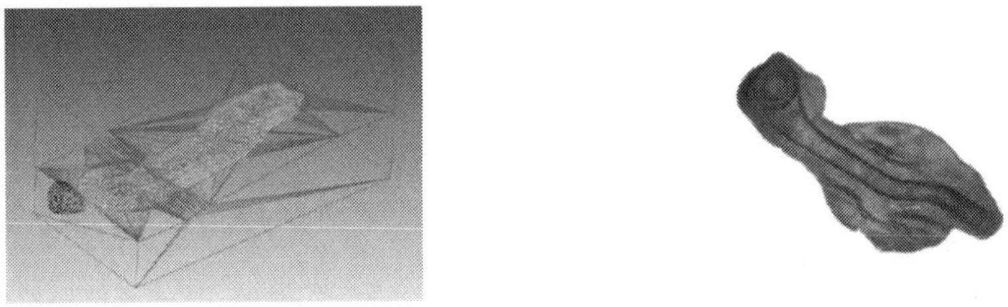

Figure 5: Initial and end configuration in final forging operation
a)CAD Deformation b) Actual Formation

To achieve this, two strokes are needed. Simulation results show that maximal contact pressures in final forging operation are lower than 900 MPa (figure.5a & 5b). Also, die filling has got completed. Based upon the above facts it can be concluded that by the implementation of optimized manufacturing of Socket Forging component of high quality can be obtained. The die stressing is significantly lower than allowed.

Conducted numerical simulation of socket forging produced by Hot Forging process has been verified in practical industrial environment, in forging plant "Punch Ratna Fasteners Pvt Ltd". Friction Drop Hammer is used in hot forging process was Rattan Hammers with capacity of 1.25 Ton, weight of the top die and Bottom Die was 3150 kg. Applied lubrication: Friction Release Lubricant oil. In figure.6, 3D model of both top and bottom die of the forging process, created using Siemens NX Cad software with all the multi stages optimized improvements getting imposed in the die design

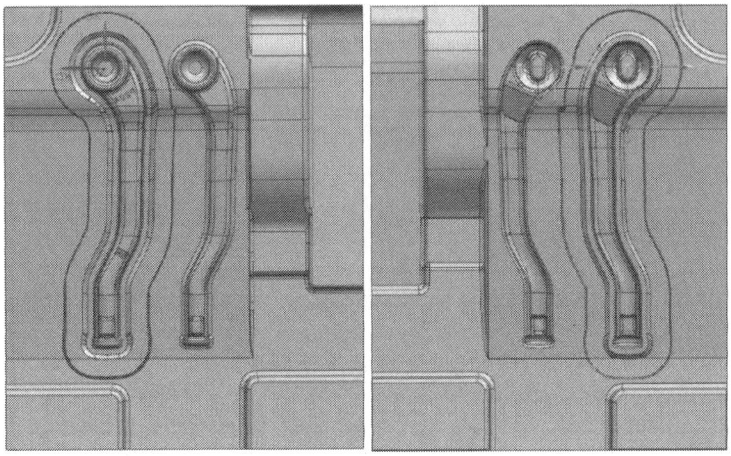

Figure 6: CAD version of the bottom and upper die

Forging parameters were identical with those applied in the numerical simulation. Results obtained in practical forging experiment confirm simulation results, i.e. complete die filling and relatively low contact pressure between the die and forging component.

IOP Conf. Series: Materials Science and Engineering **992** (2020) 012007 doi:10.1088/1757-899X/992/1/012007

3. Conclusion:

Hot forging technology is indispensable process for producing high quality parts for different applications. This technology is applied for the manufacture of automotive components, turbine discs, gears, bolts, and structural components for machinery, railroad and other transportation equipment. Due to its comparative advantages, hot forging finds wide spread implementation in metal industry all over the world.

Effective and qualitative design of forging technology requires application of modern informational technologies in all stages, starting form work piece design, multi –stage forging design, optimization of the process, die design, etc. Numerical simulation of the forging process makes possible to perform optimization of forging phases, die wear, material structure, elastic deformation of die and machine, essential for the precision of forgings.

Current paper enlightens the optimization and process design (forging multi-stage) of hot forging process for the production of Socket Forging. The aim was to determine all relevant parameters which result in correct forged part (complete die filling) and low die stressing.

Results obtained in numerical simulation were verified in practical industrial environment. By inspection of final forging component it has been concluded that they possess high quality and dimension accuracy. Also, those forgings are free of surface cracks, material overlapping and insufficient die filling.

Acknowledgement: we sincere thanks to M/s Punch Ratna Fasteners Pvt.Ltd., Chennai, has given permission to carry out experimental work. Mr. Pradeep Dayal, Plant Head, technically supported the entire fabrication work.

References:

[1] Bednarek S., Lukaszek-Solek, A., & Sinczak, J. 2008. Analysis of strain and stress in the lower forging limit of Ti-6Al-2Mo-2Cr titanium alloy. *Archives of Civil and Mechanical Engineering*, **8(2),** pp.13-20.

[2] Bruchon J., Digonnet H., & Coupez T. 2009. Using a signed distance function for the simulation of metal forming processes: Formulation of the contact condition and mesh adaptation. From a Lagrangian approach to an Eulerian approach. *International journal for numerical methods in engineering*, **78(8),** pp.980-1008.

[3[Choi D. S., & Kang H. 2020. Finite Element Analysis of Multistage Hot Forging Process during Mold Cooling. *Journal of the Korean Society of Manufacturing Process Engineers*, **19(5),**pp.75-81.

[4] Gotoh M., Sun Z G., & Katsuda M. 1997. Numerical simulation of multistage forging process: 2nd report, application to reduction of stages in a multistage forging process. *Transactions of the Japan Society of Mechanical Engineers, Series C*, **63(605)**, pp.231-238.

[5] Hong L., Sun Y., & Cao C. 2020. Motor synchronisation control for multistage hot die forging press feed manipulator based on BP-PID controller. *International Journal of Computing Science and Mathematics*, **11(4),** pp.347-356.

[6] Jayaprakash J., Thinakaran N., & Sathish T. 2014. JADE implementation of Multi-agent based Inventory routing problem. *International Journal of Applied Engineering Research*, **9(24),** pp.28785-28792.

[7] Kong L., & He M. 2019. Design and research of UT inspection tool in flaw inspection of large-size forging shaft parts. *In IOP Conference Series: Materials Science and Engineering* , **631(4),** pp. 042020.

IOP Conf. Series: Materials Science and Engineering **992** (2020) 012007 doi:10.1088/1757-899X/992/1/012007

[8] Kotous J., Kubec V., Duchek M., & Studecký T. 2020. Optimization of workability technological testing for open-die forging. *In IOP Conference Series: Materials Science and Engineering*,**723(1)**, pp. 012015.

[9] Megharaj, C. E., Nagaraj, P. M., & Pasha, K. J. (2016, September). Design and analysis of a forging die for manufacturing of multiple connecting rods. In IOP Conference Series: Materials Science and Engineering (Vol. 149, No. 1, p. 012145.

[10] Niu J T., Li H T., Meng X D., & Karjaleinen P. 2009. Numerical simulation methods and their application in material hot-working. *Acta Metallurgica Sinica*, **13(2),** pp.494-501.

[11] Obiko J., & Mwema, F. 2020. Numerical simulation on the effect of sample geometry size on the metal flow behaviour during the forging process. *IOP SciNotes*, **1(1)**, pp.014401.

[12] Ou H., Lan J., Armstrong C G., Price A., Walløe S J., & Ward M J. 2006. Reduction in post forging errors for aerofoil forging using finite element simulation and optimization. *Modelling and Simulation in Materials Science and Engineering,* **14(2),** pp.179.

[13] Shi P., Xia X., Zhou J., & Xiao P. 2019. Simulation Analysis of Forging Process of Automobile Aluminum Alloy Swing Arm. *In IOP Conference Series: Materials Science and Engineering*, **562(1),** p. 012139.

[14] Subrammanian A., Jabaraj D. B., & Jayaprakash J. 2018. Multi-objective Optimization of Resistance Spot Welding of AISI 409M Ferritic Stainless Steel. *Journal of Scientific and Industrial Research*, **77(1),** pp.271-275.

[15] Subrammanian A., Jabaraj D B., Jayaprakash J., & Raja V B. 2016. Mechanical Properties and Phase Transformations in Resistance Spot Welded Dissimilar Joints of AISI409M/AISI301 Steel. *Indian Journal of Science and Technology*, **9(41)**, pp.1-8.

[16] Zhang H., Pan A., Zhang Y., Zhang Z., & Liu J. 2019. Research on the Shunting Precision Forging of Ribbed Plate Gear Blank Based on DEFORM-3D Finite Element Analysis Under the Framework of New Engineering. *In Journal of Physics: Conference Series*, **1302(4)** pp. 042029.

IOP Conf. Series: Materials Science and Engineering **992** (2020) 012008 doi:10.1088/1757-899X/992/1/012008

Use of animal fat methyl ester blend as a fuel in a CI engine

G.Dinesh Babu and Shaafi .T

Department of Mechanical Engineering, Saveetha School of Engineering, Saveetha Institute of Medical and Technical Sciences, Chennai, India.

E-mail : shaafit.sse@saveetha.com

Abstract. A Direct Injection Simpson make (S217). Diesel engine examined for the characteristics of performance and emission fuelled with Diesel, methyl ester of waste animal fat (B100) and Diesel – methyl ester of waste animal fat (B20) blends. At constant engine speed, the parameters of emission and performance were determined for different loading conditions at constant engine speed. The peak cylinder pressure registered as 90 bar, 90.48 bar, and 89.93 bar respectively for Diesel, B100 and B20. Unburned HC and CO emissions remained declined in manner for B100 and B20 when related to diesel fuel.

Keywords: Alternative fuel, animal fat biodiesel, performance, emission.

1. Introduction

Nowadays the society highly depends on petroleum products for its day today activities. However several environmental problems are raising such as NOx, carbon dioxide etc. Diesel engines are used in almost all the fields due to high power and low operating cost. Around 30% of the global energy demand rise slowly enlarged between the period of today and 2040 [1]. Hence, there is an emergency to develop alternative fuels to diminish the need on conventional fuel. The world energy supply heavily depend on the derived fuels from the crude oil out of which 90 % is utilized for transportation and energy generation.

In the present condition, crude oil reserves depleted in a very fast manner due to the faster rate of consumption. Therefore, this scenario started and continued the interest in identifying renewable alternative fuels. Diesel engines use vegetable oils in three forms namely 1.neat or pure vegetable oils 2. Vegetable oils blends and diesel fuel, and 3. Trans-esterified vegetable oils. The first and second forms have problems associated with the long term performance of diesel engines for the reason that higher viscosity of fuel. But the esters of vegetable oils have lower viscosities than the neat or blended vegetable oil fuels and hence the viscosity related problems like clog engine components such as fuel injectors are greatly reduced. The most promising applications is esters of vegetable (edible and non-edible) used as a diesel engine fuels. Methyl, ethyl and butyl esters are prepared by the method of trans-esterification of vegetable oils.

Currently, the well-known method of biodiesel usage is blending it with conventional diesel fuel. Biodiesel (or biofuel) refers to animal fat or containing of long chain alkyl esters of vegetable oil fuel. Normally diesel, mixed with Biodiesel or biodiesel alone used as a diesel engine fuel. Now a days, in

IOP Conf. Series: Materials Science and Engineering **992** (2020) 012008 doi:10.1088/1757-899X/992/1/012008

India, most suitable feed stock for biodiesel is non-edible oil, for the reason that the requirement for the edible oil go beyond the domestic supply. Shahir et al [2] go through the usage of biodiesel advantage, when it is compared to the neat diesel. Lapuerta et al.[3] evaluated the diesel engine performance and emissions running on 1.waste animal fat biodiesel and 2. 50% animal fat & 50% soybean oil. They stated that biodiesel fuels reduces the HC emissions, opacity of smoke, and PM. NOx was increased for 50:50 soybean oil and animal fat but decreased in manner with pure animal fat. Oner and Altun [4] reported that, methyl ester of animal tallow reduces the emissions of CO, NOx, smoke opacity and SO2. Velmurugan and Shaafi [5] was conducted an experiment on a Simpsons make DI compression ignition engine with 21 kW power at 1500 rpm when running on Diesel, B20 and B100. The engine characteristics were carefully measured at various engine loads from Zero to 100 percentage. The following conclusion were arrived as B20 blend thermal efficiency was lower when compared to other two fuels and the exhaust gas temperatures are high for B20 blend compared to other fuels used it, owing to the more amount of fuel was burnt. Lin and Li [6] studied the mixed marine fish (discarded species) were used to produce the biodiesel. They testified that, biodiesel of fish-oil has greater NOx, smoke opacity and a lower bsfc rate, CO emission when compared with WCO biodiesel. Ivana et.al [7] reviewed the different methods adopted for production of biodiesel and they presented the possibilities of exploitation of waste animal fats. Awad et.al [8] found the correlation between the parameters of combustion parameters and properties of fuels. Preheated condition of animal fat used as a fuel in a CI engine without any deterioration in performance of the engine.[9]. Mugilan and Shaafi [10] conducted an experiment on a diesel engine consists of twin cylinder was running using palm oil biodiesel and diesel-palm oil biodiesel blend. The loads of the engine are varied to 25%, 50%, 75% and 100% and the performance and emissions of exhaust were measured accordingly. They noticed that emissions of CO and HC were low and at the same time, higher NOX emissions when compared to diesel fuel. Diesel with 20% methyl ester of palm oil blend was viable as an alternate fuel for diesel engine. Shahir et.al [11] found that the use of enormous amount of animal fat biodiesel would solve the pollution created from the waste of meat.

Based on the literature survey, limited work has been concentrated on animal fats biodiesel as an alternate fuel in a CI engine. In this paper, a Simpson make (S217). DI Diesel engine examined for the performance and emission characteristics, fuelled with Diesel, methyl ester of waste animal fat (B100) and Diesel – methyl ester of waste animal fat (B20) blends.

2. Materials and Methods

2.1. Description of the engine

A Direct Injection Simpson make (S217). Diesel engine was examined for the characteristics of performance and emission fueled with Diesel, waste animal fat methyl ester (B100) and Diesel -waste animal fat methyl Ester (B20) blends. The Figure.1.shows the photographic view of the diesel engine with loading device and the specifications of the engine are shown in table.1.

Model	SIMPSON S 217
Capacity	21 kW
Bore	91.44 mm
Stroke	127 mm
No.of cylinders	2
Compression ratio	18.5 : 1
Combustion system	Direct Injection

Figure.1 Diesel Engine with loading device

2.2. Fuel properties

Table 2. Showing the diesel, waste animal fat methyl ester and its blend properties of the tested fuel. Biodiesel density is higher in order and the heating value is lower in order when compared to the diesel. Figure. 2 shows the prepared methyl ester of waste animal fat.

Figure. 2 Prepared waste animal fat methyl ester

IOP Conf. Series: Materials Science and Engineering **992** (2020) 012008 doi:10.1088/1757-899X/992/1/012008

Table 2 Properties of the tested fuel

Fuel	Diesel	B100	B20
Density kg/m^3	830	829	814
Kinematic viscosity	1.96	2.93	2.57
Calorific value kJ/kg	47000	36835	37418
Cetane Index	46	53	55

3. Results and Discussion

3.1 Brake thermal efficiency

Figure.3.shows the variation of brake thermal efficiency at various loading conditions. Owing to the higher viscosity of B100 and B20 (related to conventional diesel) fuels are leads to 1.dwindled atomization 2.poor fuel evaporization and incomplete combustion. Hence, the efficiency of brake thermal for B100 and B20 are lower at higher loads than the diesel fuel.

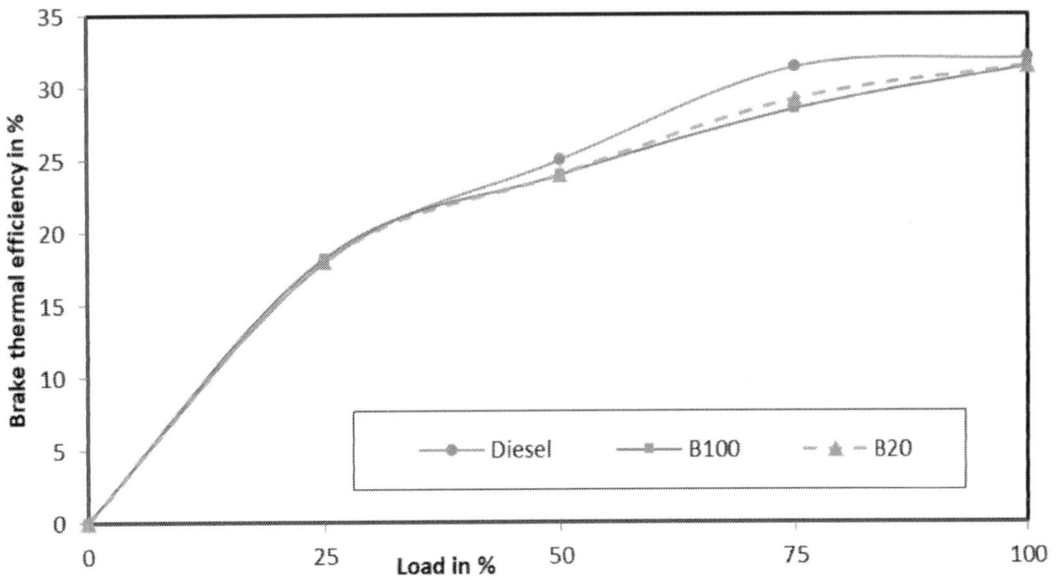

Figure 3. Brake thermal efficiency at various load.

3.2 Cylinder Pressure

The cylinder pressure variation is shown in Figure.4. A slightly reduced peak pressure recorded for B20 and B100 related with diesel. The pressure rise rate is also highest for diesel fuel and lowest for B20. Higher peak pressure occurs between the crank angles from 0 to 10 degree.

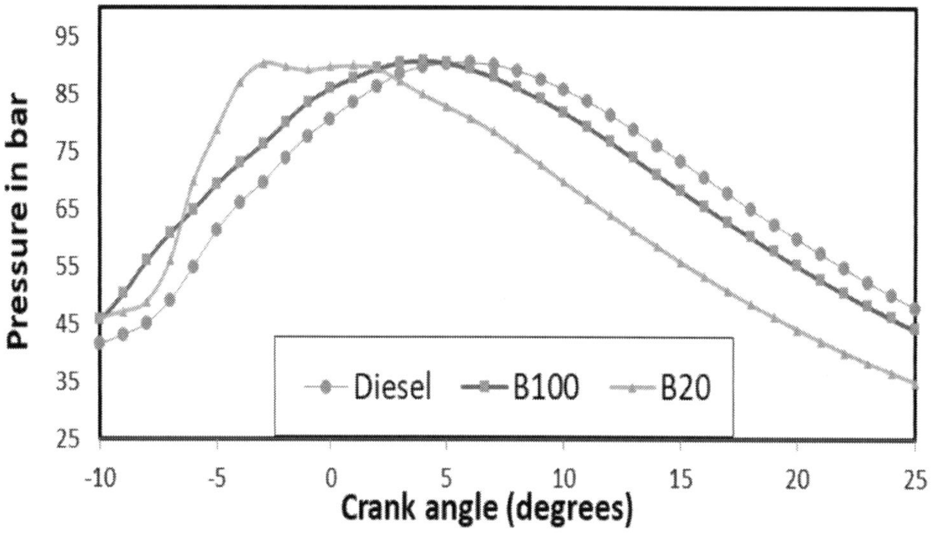

Figure 4. Variation of Cylinder Pressure

3.3 Exhaust temperature

The variations of exhaust gas temperature of the tested fuels are shown in Figure 5.The EGT of B20 fuel is lower than tested the other two fuels at all the loads. Variations of the EGT indicates the kind of fuel used and brake loads have the effect on the temperature of the exhaust. The maximum EGT is recorded 530°C at full load for diesel.

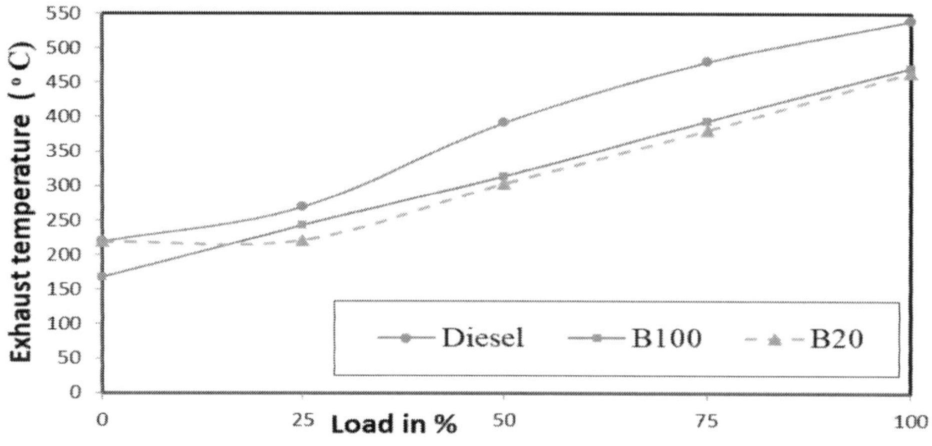

Figure 5. Exhaust gas temperature variation

3.4 Co emission

Figure 6.shows the emissions of CO variation for diesel, B100 and its blends. Due to higher viscosity the CO emission increased in trend from lower load to higher load. Up to 25% of the load there is no much difference for all the tested fuels.

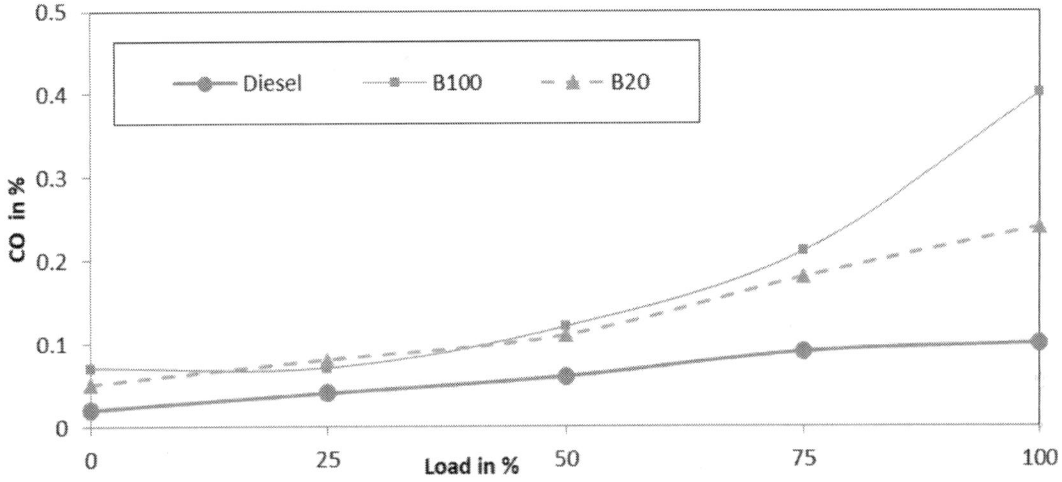

Figure 6. CO emission variation

3.5 UBHC emission

Figure 7. shows HC emissions variations for diesel, biodiesel and its blend. There is rise in hydro carbon emissions for B20 and B100 are higher at loads from 75% to 100%. Elevated combustion temperature and oxygen content availability in the biodiesel which promote the oxidation of hydrocarbon emissions. At a lower load, the B20 and B100 are having lower HC emissions compared to diesel.

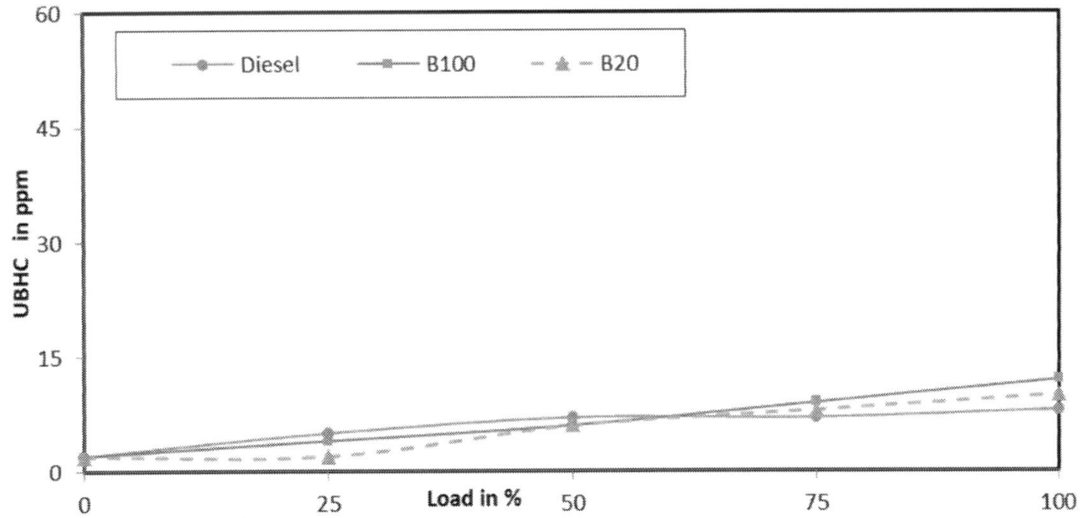

Figure 7.UBHC emission variation

3.6 NOX Emission

NOx emission variation for diesel, B100 and B20 are shown in Figure.8. Cetane number influences the NOx emissions for CI engines. Higher EGT and higher NOx formation are occurs owing to the lower value of cetane number causes the increase in ignition delay for B20 and B100, but when it compared with neat diesel the magnitude was low.

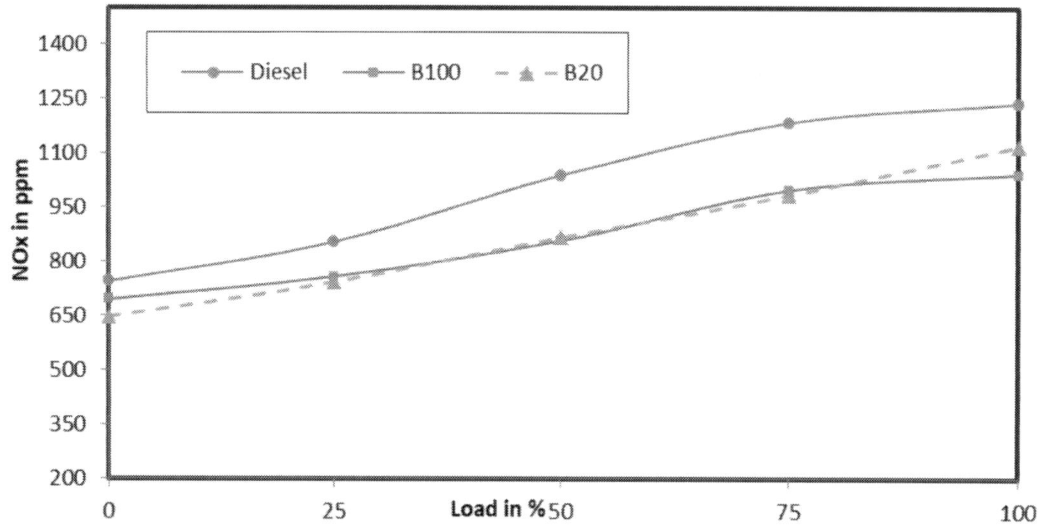

Figure 8.NOx emission variation

3.7 Smoke emission

Figure.9. shows the density of smoke variation with diesel, biodiesel and its blend. The density of the smoke decreases for B20 and B100, owing to the availability of lesser carbon content in the biodiesel and its blend.

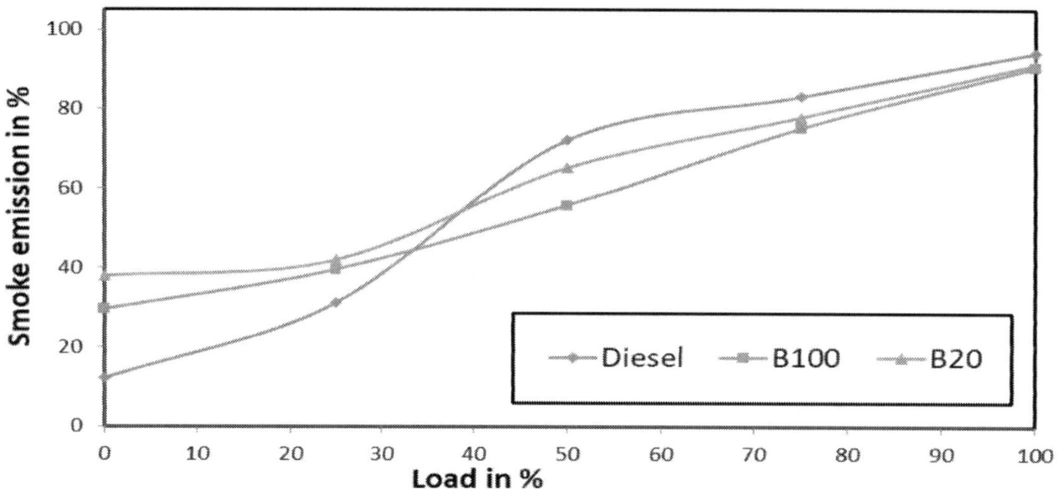

Figure 9. Smoke emission variation

IOP Conf. Series: Materials Science and Engineering **992** (2020) 012008 doi:10.1088/1757-899X/992/1/012008

4. Conclusion

In this work, Performance and emissions of a DI compression ignition engine running on B100, B20 were examined and compared with diesel.

The conclusions were summarized as follows:

- ✓ due to diminished atomization and poor fuel vaporization, B100 and B20 brake thermal efficiencies are lower at higher loads than the diesel fuel.
- ✓ B20 blend fuel, exhaust temperature is lower compared to the tested other two fuels at all the engine load.
- ✓ The density of the smoke decreases for B20 and B100. The availability of low carbon content in the biodiesel noticed that delivered very low smoke density emission.

References

[1] www.iea.org.

[2] shahir.V.K, Jawahar C.P and Suresh PR. "Comparative study of diesel and biodiesel on CI engine with emphasis to emissions - A review" Renewable and Sustainable Energy Reviews 45, 686–697, 2015.

[3] Lapuerta, M., Rodrıguez-Fernandez, J., Oliva, F. and Canoira, L., "Biodiesel from Low-Grade Animal Fats: Diesel Engine Performance and Emissions", Energy and Fuels, 23, 121-129, 2009.

[4] Oner, C. and Altun, S "Biodiesel Production from Inedible animal tallow and an experimental investigation of its use as alternative fuel in a Direct Injection Diesel Engine", Applied Energy, 86, 2114-2120, 2009.

[5] Velmurugan. K, Shaafi. T "Performance and emission characteristics of a twin cylinder di diesel engine running on diesel - waste cooking methyl ester blend" International Journal of Mechanical and Production Engineering Research and Development, vol.9 issue 6, pp.183-190.

[6] Lin, C.Y. and Li, R.J., "Engine Performance and Emission Characteristics of Marine fish-oil biodiesel produced from the discarded Parts of Marine Fish", Fuel Processing Technology, 90 883-888, 2009.

[7] Ivana B.Banković-Ilić , Ivan J.Stojković , Olivera S.Stamenković, Vlada B.Veljkovic ,Yung-Tse Hung "Waste animal fats as feed stocks for biodiesel production" Renewable and Sustainable Energy Reviews 32 ,238–254, 2014.

[8] Sary Awad, Khaled Loubar, Mohand Tazerout "Experimental investigation on the combustion, performance and pollutant emissions of biodiesel from animal fat residues on a direct injection diesel engine" Energy, 69, 826-836,2014.

[9] M.Senthil Kumar, A.Kerihuel, J. Bellettre, M. Tazerout "Technical note Experimental investigations on the use of preheated animal fat as fuel in a compression ignition engine" Renewable Energy 30,1443–1456, 2005.

[10] Mugilan. M, Shaafi. T "Effect of Palm 0il Biodiesel-Diesel blended fuels on Engine Performance and Emission Characteristics" International Journal of Mechanical and Production Engineering Research and Development, vol.9 issue 6, pp. 943 -950.

[11] V.K.Shahir, C.P.Jawahar P.R.Suresh, and V.Vinod "Experimental Investigation on Performance and Emission Characteristics of a Common Rail Direct Injection Engine Using Animal Fat Biodiesel Blends" Energy Procedia. 117, 283-290, 2017.

IOP Conf. Series: Materials Science and Engineering **992** (2020) 012009 doi:10.1088/1757-899X/992/1/012009

Experimental investigation on the mechanical properties of wood sawdust and plaster of paris reinforced composite

Avinash Singh and A Muniappan

Institute of Mechanical Engineering, Saveetha School of Engineering, Saveetha Institute of Medical and Technical Sciences, Thandalam, Chennai, India

E-mail: avinashsingh.singh793@gmail.com

Abstract. In this work the wood saw dust and plaster of paris have been reinforced to form a new composite material. The materials which have been used are wood saw dust and plaster of paris. Wood saw dust is a derivative of several operations on wood such as drilling, routing, sanding, milling, sawing, planning etc. and plaster of paris is made from gypsum. Gypsum is heated at 373 K in a furnace to lose some water content and thus POP (plaster of paris) is formed. Both materials are mixed together with the help of water to form a paste and are freezed at 0°C for 24 hours. And three different samples are made by varying the ratio of sawdust and plaster of paris. The samples are obtained in the 15cm x 15cm x 3.5cm dimensions by using a plastic box and is given in the laboratory for testing various properties such as density, ash content, hardness, compression load, hardness at -40°C etc. Three different samples are in the ratios (WSD: POP): 3:1, 1:1, 1:3 are obtained. Out of three the third sample gives the best result and can be used as a construction material in cold places.

Keywords. Composite, wood sawdust, plaster of paris, properties, hardness

1. Introduction

Sawdust is a waste product formed by various operations done on wood such as milling, cutting, planing, sawing, routing etc., it has no practical use. However it has been used several times to form a new composite which can be used in construction sector [1], [5], [6]. Plastic waste is one of the major problems nowadays and is responsible for environmental pollution to a great extent. Thus it has been blended with wood sawdust so as to reduce plastic accumulation and reduce trees cutting [3], [8], [10]. Wood sawdust is also mixed with polypropylene to enhance the properties of wood sawdust and polypropylene such as thermal stability, water absorption etc. so that it can be used in furniture industry, automobile industry and various other industries [2], [4], [9]. Wood sawdust is also mixed with the waste of plant to form chipboards and other items based on dispersed wood [7]. In this project wood saw dust (WSD) and plaster of paris (POP) have been reinforced together to get a new composite material. Plaster of paris has not been used with wood sawdust before to make any construction material. In this paper a total of four experiments have been carried out to arrive at why POP should be used as binding agent with WSD and in last experiments three samples have been prepared with different WSD to POP ratios in order to get the best composition of WSD and POP which gives the best desirable properties.

2. Materials used

Materials which are used in this composite are wood sawdust and plaster of paris. Wood sawdust is formed as a waste product by various operations on wood such as sanding, cutting, milling, drilling etc. Plaster of paris is formed by heating gypsum at 373 K in a furnace. $CaSO_4.2H_2O$ (gypsum) → $CaSO_4.1/2H_2O$ (plaster of paris) + $3/2H_2O$. Here $CaSO_4.1/2H_2O$ is plaster of paris and its chemical name is calcium sulphate hemihydrate. Its setting time is 15-20 minutes, density is 780-800 kg/m^3, transverse strength is 15-20 kg/m^2, compressive strength is 1-3 N/mm^2, whiteness is 80-85, fineness is 99, shelf life is six months from date of manufacturing, manufacturing is 'in house'. Plaster of paris is used in plastering the fractured bones, used in dentistry, making toys, statues and decorative items etc. Another important material which is used in this composite is water, its density in liquid form is 1g/ml at 0°C and boiling point is 100°C.

3. Methodology

It is very simple to make WSD-POP composite, take some quantity of wood saw dust and plaster of paris and mix it well with water to obtain a semi solid paste. Obtain a specific shape with the help of container in which the paste will be kept, fix it properly with the hands so that no voids is left out and keep the container in the freezer (0°C) for 24 hours so that it gets solidified properly. After this take out the composite keep it at the room temperature and note down its temperature. After 15 minutes take out the composite from the container and give it in the laboratory for testing.

4. Experiments and observations

Experiment 1: take 30g of wood sawdust and mix it with 15g of water and keep the paste in small paper glass, allow the composite to get freezed at 0°C for 24 hours. After 24 hours take out the composite from the cup and allow it to cool down at normal room temperature. Note down its temperature after 5 minutes from the time it has been taken out. Observation: temperature of the composite after it has been taken out from the freezer (after 5 minutes) is 10°C, time for which the composite retained its shape is 120 minutes. Inference is there is a need of using a binder to give permanent shape to the composite. Experiment 2: take 30g of wood sawdust and add water in three different quantities i.e. 10g, 20g, 30g, and allow the three samples to get freezed for 24 hours at 0°C. Observation: the temperature of the three samples at room temperature is 0°C and time for which composites retain its shape (all three samples) is 120 minutes. Inference is quantity of water has no effect on the composite. Experiment 3: take 30g of wood sawdust and mix it with different other possible materials which can act as a binder and add water accordingly to get a semi solid paste. Keep the paste in freezer and allow it to get freezed for 24 hours. Take out the composite and note down its temperature and time retention for its shape (table 1).

Table 1. Observation for experiment 3.

Sawdust + binding agent	Temperature (°C)	Shape retention time
Calcium	10	30 min
Plaster of paris	10	permanent
Detergent	08	15 min
Readymade cement	10	40 min
White cement	10	45 min
Kolam powder	08	30 min

Inference is wood sawdust with pop gives the best result. Experiment 4: prepare three samples (figures 1, 2, 3) for WSD-POP composite by varying the ratios of wsd to pop i.e. 3:1, 1:1, 1:3 (table 2).

Table 2. Composition of different samples.

Parameters	Sample 1	Sample 2	Sample 3
Total mass	500g	500g	500g
WSD	375g	250g	125g
POP	125g	250g	375g
Water	1000ml	550ml	450ml

Figure 1. Sample 1: the ratio of WSD:POP by mass is 3:1.

Figure 2. Sample 2: the ratio of WSD:POP by mass is 1:1.

Figure 3. Sample 3: the ratio of WSD:POP by mass is 1:3.

Give all the samples in the laboratory for determining its density, ash content, hardness, compressive strength, hardness at low temperature (-40^{0}C), sample size is 15cm × 15cm × 3.5 cm.

5. Result and discussion

Properties of all the three samples have been noted down in table 3 and graphs have been made to do comparative study of properties of all the three samples (figures 4, 5, 6, 7, 8).

Table 3. Properties of all three samples.

Parameters	Sample 1	Sample 2	Sample 3
Density	0.59g/Cc	0.95g/Cc	1.24g/Cc
Ash content	8.6%	17.76%	37.2%
Shore 'a' hardness	NA	30,30,35,35,33	79,79,80,80,80
Condition at low temperature (-40°C), kept for 5 minutes	the sample cracks	sample becomes hard	sample becomes hard
Compressive load	NA	0.4 KN	1.6 KN
Shore 'a' hardness at -40°C kept for 5 minutes	NA	40,40,42,42,42	82,82,85,85,85

From figure 4 we can make out that as the proportion of pop increases the density of samples also increases as pop is denser than wsd. In figure 5 as the proportion of pop increases the ash content of the composite also increases. In sample 1 the proportion of pop is less that's why its hardness cannot be obtained due to its poor strength. And hardness increases as pop proportion increases in the composite (figure 6). At -40oC temperature, sample 1 cracks whereas samples 2 and 3 become harder, sample 1 cracks due to less proportion of binding agent pop, whereas sample 2 and sample 3 become harder due to their comparatively higher proportion of pop (binding agent). In figure 7, as pop proportion increases the compressive strength of the sample also increases. As pop proportion increases the sample becomes harder at low temperature (figure 8).

IOP Conf. Series: Materials Science and Engineering **992** (2020) 012009 doi:10.1088/1757-899X/992/1/012009

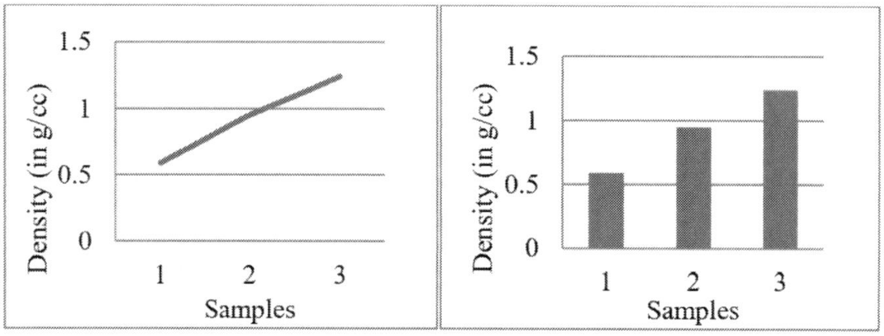

Figure 4. Samples Vs density.

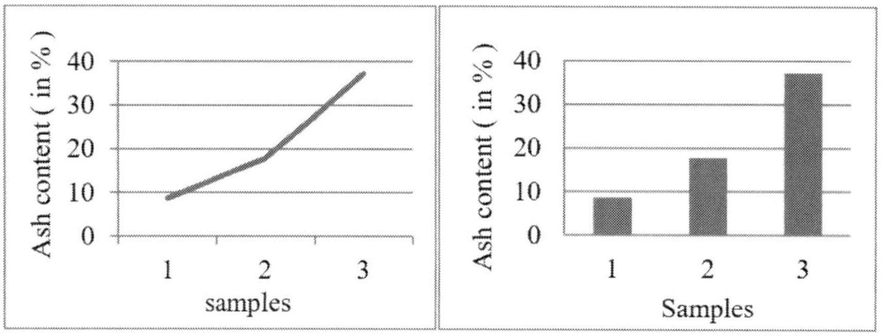

Figure 5. Samples Vs ash content.

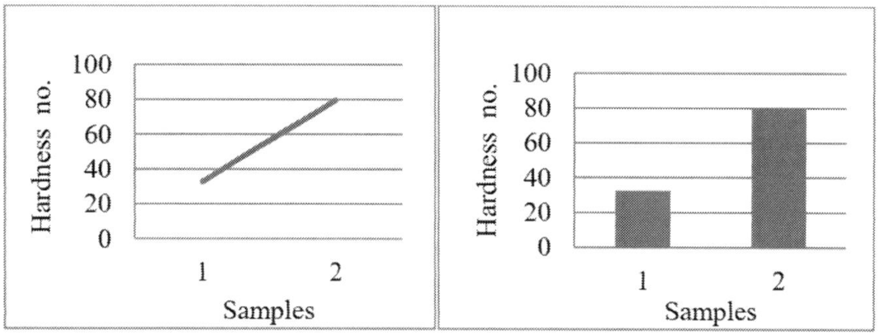

Figure 6. Samples Vs hardness number.

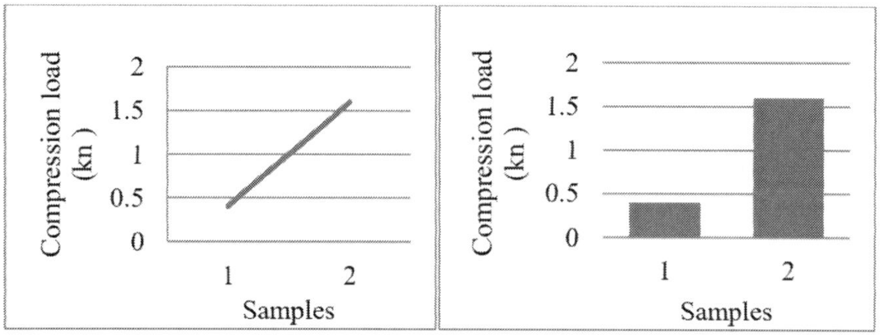

Figure 7. Samples Vs compression load.

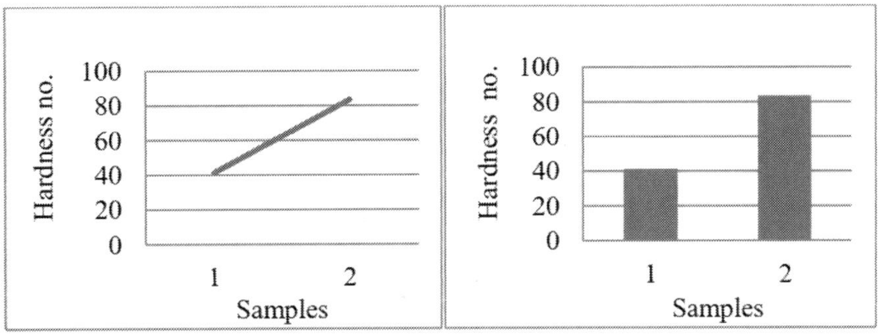

Figure 8. Samples Vs hardness number at -40°C.

6. Conclusion

From the results and discussion we can conclude that the desirable properties for the composite material (WSD-POP Composite) such as hardness, compressive strength, hardness at low temperature, increases with increase in pop proportion. Thus, the composite with wsd and pop in the ratio 1:3 gives the best result and can be used as the construction material in cold places.

Acknowledgements

I express my thanks to my project guide Mr A Muniappan, Assistant Professor at Saveetha School of Engineering, SIMATS for his support. I also extend my warm thanks to my college Saveetha School of Engineering, SIMATS for giving this opportunity and Kidao Laboratories, Chennai, India for carrying out the testing of my samples.

References

[1] Memon R P, Sam A R M, Awal A S M A and Achekzai L 2017 *Mechanical and Thermal properties of Sawdust Concrete* 23-7

[2] Fakhrul T, Mahbub R and Islam M A 2013 **vol.1** (no.1) *Properties of Wood Sawdust and Wheat Flour Reinforced Polypropylene Composites* 135-48

[3] Mwema M F, Mburu N J, Mutunga N B and Kalama F C 2015 **vol.5** (issue 12, part 4) *Development of Wood Plastic Composites* 11-7

[4] Idrus M A M M, Hamdan S, Rahman M R and Islam M S 2011 *Treated Tropical Wood Sawdust Polypropylene Polymer Composite : Mechanical and Morphological Study* 435-44

[5] Hafed S A 2017 **vol.3** (issue 2) *Study of Thermal Insulation and some Mechanical properties for Hybrid Composites (Cement-Wood Sawdust)* 212-6

[6] Awal A S M A, Mariyana A A K and Hossain M Z 2016 **vol.10** (issue 21) *Some aspects of Physical and Mechanical properties of Sawdust Concrete* 1918-23

[7] Raspopv L N and Matkovskii P E 2011 **vol.2** (no.3) *Wood–Mineral-Polymer Composites materials* 117-30

[8] Ajaj I A A and Obaidi M F A 2015 **vol.5** (no.1) *Effects of Sawdust Wood on some Mechanical properties of recycled Polyethylene Terephthalate (RPET)* 20-6

[9] Chiper A M, Nemes O, Soporan V F and Rus A R 2010 *Composite plates from Wood-Flour* 28-32

[10] Brostow W, Datashvili T, Jiang P and Miller H 2016 *Recycled HDPE reinforced with sol-gel Silica modified Wood Sawdust* 28-39

IOP Conf. Series: Materials Science and Engineering **992** (2020) 012010 doi:10.1088/1757-899X/992/1/012010

Heuristics based hybrid approach of flow shop scheduling problem

K Lokesh[1], J Jayaprakash [2]and Tibebu Alene[3]

[1,]UG Student, [2]Professor, Department of Mechanical Engineering, Saveetha School of Engineering , Saveetha Institute of Medical and Technical Sciences, Chennai, India-602105.
[3]Faculty, Department of Industrial Engineering, University of Gondar, Ethiopia.

E-mail: [1]lokeshkannan43066@gmail.com

Abstract. In recent day's furniture fabrication industries are one of the upcoming segments in developing countries. In the fast-developing economy and populations are substantial interest in house hold furniture and wooden games things. Almost around more than 100 medium and large furniture businesses are running in every district capital. Presently they are confronting additionally testing undertakings in adjusting of cycle arranging and assembling products based on customer requirements. In this paper, we have taken scheduling issues related with these enterprises. In the case company contain a set of tasks to be carried out and optimize make span with a set of machines available to perform those jobs. So as to lead important information writing study on past investigations, we proposed heuristics-based hybrid approach of flow shop scheduling problems to assess the make span. At last, we verify the outcomes of the hybrid approach with LEKIN programming.

1. Introduction

The Scheduling is a significant dynamic cycle that is utilized consistently in different assembling and administration companies. It deals with the optimization of various operational objectives. And also, it deals in allocating the resources to tasks over given period of time. One target might be minimization of assignments finished after their individual due dates, others may need to satisfy the interest inside a specific due date falling flat in which may prompt punishments, hardly any others may have the goal of limiting the quantity of late positions in the part and so forth.

In flow shop scheduling, condition of machines are thought to be set up in arrangement, the work in an occupation is separated into discrete assignments called tasks, and every activity is performed at an alternate machine. These operations should be sequenced to select the next task and scheduled to plan the start time and completion time.

Johnson [1] first introduced algorithm that can find ideal arrangement for a n-work and 2-machine scheduling problem. Palmer [2] presented a solitary emphasis technique to reach required result of minimizing the make span. Here the sequencing of the positions was done dependent which are then sorted in the decreasing order so as to obtain the optimum sequence by palmer's method which are then sorted in the decreasing order so as to obtain the optimum sequence using palmer's method This algorithm can be applied to m-machine and n-work flow shop scheduling issues [3].

The basic Johnson's computation was expanded using Campbell [4]. In contrast to the altered Johnson's calculations, it utilizes various emphasis before arriving at the final result. This Johnson's extended method is a heuristic approach usually known as Campbell, Dudek and Smith heuristics (CDS). Gupta [5] Introduced minimizing the completion time and the make span of the sequence. An alternate method is set up by him in getting the slop index, in light of which the sequencing of occupations is done in the flow shop.

Zakria [6] focused on a stochastic scheduling approach to get the required schedule of jobs. It uses a stochastic environment along with a single machine to carry out the target of minimizing the lateness for the completion of the job. This included a group of jobs with different due dates assigned to them, where we need to find the lateness value of each of the jobs.

Modrak [7] illustrated the comparison between the various heuristic's algorithms based on the output value of the make span. Wu [8] acquired the concept of learning effect for the development of a model to minimize the completion time. The total completion time in various flow shop environments is minimized as required by this method by introducing the concept of learning. This paper shows some of the essential methods of solving flow shop scheduling problems focusing on finding the most efficient way of scheduling furniture manufacturing industry which is a flow shop environment that provides different types of goods (namely, bed, kitchen cabinets, drawers, dining tables, and doors.) with the help of Palmer's heuristics algorithm with LENKIN software, Comparison was also done between those heuristics algorithms to obtain an optimal make span.

Zhang [9] reviewed job shop problems related to Inudstry4.0 Varela [10] compared different configuration of parallel scheduling problems. Toader [11] reviewed flexible scheduling problem in various production systems. Simeonovova [12], [13] and Azemi [14] focused important between planning and scheduling problems. Abide [15] discussed machine repair model. Piairo,[16], Lia [17] implemented customer centered approach in scheduling problems. May [18] increased the search efficiency by genetic algorithm in job shop problems. Duplakova,[19] proposed simulation software to optimize production scheduling problems. Coman [20], [21], Irman [22], [23] A optimized make span in job shop problems. Balog [24] proposed LENKIN simulation software to optimize time frame in manufacturing scheduling problems. Jayaprakash [25] focused Multi agent based algorithm.

2. Methodology

The Flow shop scheduling problem is evaluated for improving performance of the machines and to obtain the optimal processing time. Palmer's Heuristic algorithms are used for analysis sample case problems and Integrate Palmer's result as input to LENKIN software and compare the results.

IOP Conf. Series: Materials Science and Engineering **992** (2020) 012010 doi:10.1088/1757-899X/992/1/012010

2.1 Palmer's Algorithm:

This technique will attempt to discover a weighted total for every one of these positions. In this way, we will offer loads to every one of these machines and afterward we attempt and discover the weighted whole of each positions as shown in equation 1. Pseudo code of Palmer's algorithm is given below:

t = 0, n, m; // start scheduling with m machine, n jobs //
Assign P_w (t); // Assign weights to each machine //

$$S(j) = -\sum_{f=1}^{k}(k-(2*t-1))*o(j,t) \quad \ldots\ldots\ldots \tag{1}$$

Eval P_j (t) = P_w (t).Tj; // Evaluate weights for each job //
Sort Pj(t); // Sort all jobs based on weights in descending order //
Cal P_m (t) //Calculate make span based on sequence//

2.2 Lekin software system

Lekin software can be accommodating various machine scheduling environments such as single machine scheduling problem, parallel machine scheduling problem, flow shop & Job shop scheduling problems and Flexible flow shop & job shop scheduling problems.. Firstly the user has to select machine environment and enter all the machine data and job data manually. After finishing the necessary information, the user should select scheduling method from the tool bar. The two types of scheduling methods are: rule (which consist ATCS, EDD, MS, FCFS, LPT, SPT, WSPT and CR scheduling tools) and heuristic (which consists general SB routine, shifting bottleneck/ sum (wT), shifting bottleneck/ Tmax, and SB-LS (FFS decomposition)). Here for this research heuristic scheduling method is applied.

3. Results and Discussion

Scheduling is a decision-making process to assign specific job to specific machine in many manufacturing and services industries. So that we can optimize make span.The model of the furniture industry is n/m/F/M where; n = the number of jobs, T = the number of machines, F = flow shop problem, M = make span In this work, we have taken 5 jobs and 5 machines. The machines and jobs performed are listed on the following table 1.

Table 1: Operation time matrix for 5 jobs and 5 Machines.

	T1	T2	T3	T4	T5
J1	30	25	85	40	60
J2	20	150	165	60	480
J3	35	120	90	80	180
J4	60	180	240	240	340
J5	40	75	110	90	150

J1= dressing table, J2= bed, J3=drawer, J4=kitchen cabinet, J5= door.
T1= Wood cutting machine, T2= Rotter machine, T3= Fitting machine, T4=Sander machine, T5= Painting machine.

3.1 Heuristic algorithms

3.1.1 Using Palmer's algorithm

Implement this algorithm, we can find out a weightage sum of each job.

Step 1: In this case flow shop scheduling problem having 5 machines and 5 jobs.

Step 2: Based on important of machine utilization assign weights to each machine. weight T1 = -3, machine weight T2 = -2, machine weight (T3) = 0, machine weight (T4) = +2, machine weight (T5) = +3 as shown in table 2. Positive sign means most wanted and negative means not that much.

Table 2: Different weighted factor.

	T1	T2	T3	T4	T5
J1	30	25	85	40	60
J2	20	150	165	60	480
J3	35	120	90	80	180
J4	60	180	240	200	340
J5	40	75	110	90	150
	-3	-2	0	+2	+3

Step 3: this step we have to calculate the weight of each job, and then multiplying the weights to the processing times. The time is in minutes /some units

W (J1) = (-3*30) + (-2*25) + (0*85) + (2*40) + (3*60)=120;
W (J2) = (-3*20) + (-2*150) + (0*165) + (2*60) + (3*480) = 1200
W (J3) = (-3*35) + (-2*120) + (0*90)
+(2*80) +(3*180) = 355
W (J4) = (-3*60) + (-2*180) + (0*240) +(2*200) +(3*340) = 880
W (J5) = (-3*40) + (-2*75) + (0*110) + (2*90) + (3*150) = 360

Step 4: Sorting of calculated job weights in descending order. In decreasing order sequence, the jobs become :**J2-J4-J5-J3-J1**

Step 5: Formulating sequence of jobs based on the previous step done in Step 4.
The sequence of the job in tabular form can be expressed in table 3.

Table 3: Optimal sequence with weight

machine job	T1	T2	T3	T4	T5	weight
J2	20	150	165	60	480	1200
J4	60	180	240	200	340	880
J5	40	75	110	90	150	360
J3	35	120	90	80	180	355
J1	30	25	85	40	60	120

Step 6: finally calculate make span of the given jobs with the given machines in optimal value as followed in table 4

Table 4: Final make span using Palmer's algorithm

Machine Job	T1	T2	T3	T4	T5
J2	20	170	335	395	875
J4	80	350	590	790	1215
J5	120	425	700	880	1365
J3	155	545	790	960	1545
J1	185	570	875	1000	**1605**

Make span associated with this sequence is 1605 minutes in table 4.

3.1.2. Using LEKIN software system

In order to run the LEKIN software the required data are number of machines, number of jobs and then enter processing time of each jobs in each machine.

The total processing time is displayed by default when the above data is entered to the system correctly. The following figure 1 gives detail expression for the above points.

The LEKIN software below shows the completion time total completion of each jobs using heuristics scheduling by shifting bottleneck/ Tmax. The maximum completion time calculated is 1499 minutes. Total processing time 185for machine one, 550 for machine two, 690 for machine three, 510 for machine four and 1210 for machine five; 240 for job one, 875 for job two, 505 for job three, 1060 for job four and 465 for job five. Other information like beginning, ending, total and weighting time also expressed in this method. The detail optimal results using Lenkin software as shown in figure 2

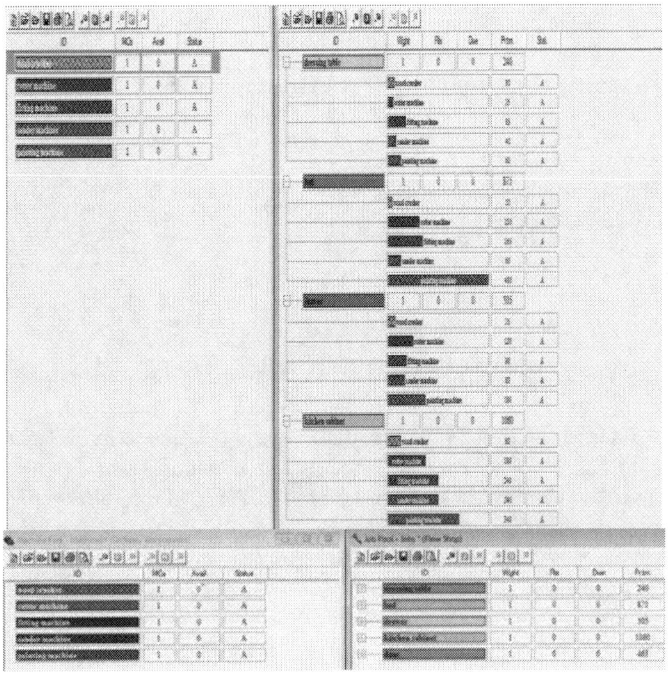

Figure 1: Job allotment to each sequence using Lekin software

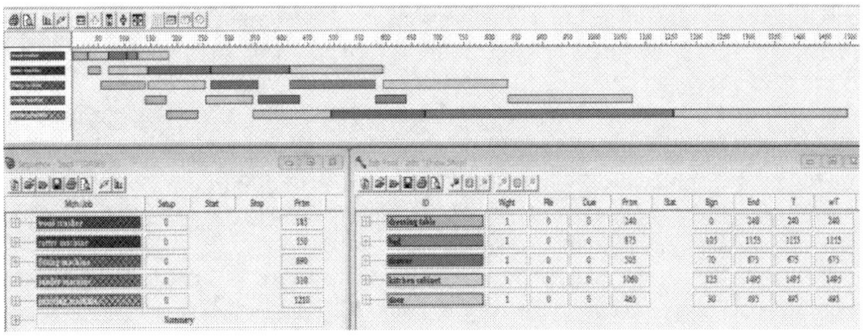

Figure 2: Optimal result using Lekin software

3.2 Comparison of heuristic techniques:

Table 5: comparison result with different algorithm

Heuristic algorithm	**Optimum value in minutes**
Palmer's algorithm	**1605**
Hybrid Approach (Shifting bottleneck/Tmax)	**1499**

4. Conclusion

This paper we have discussed Palmer's algorithm to minimize the make span and to enhance the performance resources. Comparison was done for the best heuristic's algorithm for minimization. From the results of a 5 job 5 machine problems, the make span calculated is 1605 minutes for Palmer's algorithm, and 1499 minutes by Shifting bottleneck/ Tmax using Hybrid approach using LENKIN software as shown in table 5. The above sequence emphasizes that, the total completion time for the jobs to be performed on those five machines is minimized.

References

[1] Johnson. 1954. Optimal two and three stage production schedules with setup times included. A *Journal Dedicated to Advances I n Operations and Logistics Research,* pp.61-68.

[2] Palmer. 1965. Sequencing Jobs through a Multi-Stage Process in the Minimum Total Time- A Quick Method of Obtaining a Near Optimum. *Journal of operational research society*, pp.101-107.

[3] Pinedo, L. 2012. Scheduling theory, Algorithms and Systems. New York : Springer

[4] Campbell E. 1970. A heuristic algorithm for the n job, m machine sequencing problem. *Journal of management science*, pp.630-637.

[5] Gupta. 1971. A Functional Heuristic Algorithm for the Flow shop Scheduling Problem. Journal Of Operation Research Society, pp.39-47

[6] Zakria E. 2010. Stochastic Scheduling to Minimize Expected Lateness in Multiple Identical Machines. *International Journal of Mechanical, Aerospace, Industrial, Mechatronic and Manufacturing Engineering*, pp.103-115.

[7] Modrak E. 2013. Performance measurement of selected heuristic algorithms for solving scheduling problems. *European operation research*, pp.158-183.

[8] Wu L. 2009. A note on the total completion time problem in a permutation flow shop with a learning effect. *European Journal of Operational Research*, **192(1)**, pp.343-347.

[9] Zhang J., Ding G., Zou Y., Qin S., & Fu J. 2019. Review of job shop scheduling research and its new perspectives under Industry 4.0. *Journal of Intelligent Manufacturing*, **30(4)**, pp.1809-1830.

[10] Varela M., Trojanowska J., Carmo-Silva S., Costa N M., & Machado J. 2017. Comparative simulation study of production scheduling in the hybrid and the parallel flow, *Management and Production Engineering Review,* **8(2)**, pp.69-80

[11] Toader F. A. 2017. Production scheduling in flexible manufacturing systems: a state of the art

survey. *Journal of Electrical Engineering, Electronics, Control and Computer Science,* **3(1)**, pp.1-6.

[12] Simeonovova I., Gyurov A., &Simeonov S. 2015.Modern methods for manufacturing planning and scheduling. *MM Science Journal.* pp.635-639.

[13] Syarif, A., Wamiliana, W., Lumbanraja, P., & Gen, M. (2020, May). Study on genetic algorithm (GA) approaches for solving Flow Shop Scheduling Problem (FSSP). *In IOP Conference Series: Materials Science and Engineering,* **857(1)**, pp. 012009.

[14] Azemi F., Simunovic G., Lujic R., &Tokody D. 2019. Intelligent Computer-Aided resource planning and scheduling of machining operation. *Procedia Manufacturing,* Vol:**32**, pp.331-338.

[15] Abide A., Jayaprakash J., Haile B., &Geremew S. 2018. Machine Repair Problem with Preventive Maintenance and Multi Criteria Prioritization of Machines. *International Conference on Advances of Science and Technology*, pp. 83-92.

[16] Piairo J., Madureira A., Pereira J.., & Pereira I. 2013. A user-centered interface for scheduling problem definition. *In Advances in Information Systems and Technologies*. pp. 1063-1073.

[17] Lia E. 2009. Efficient composite heuristics for total flowtime minimization in permutation flow shops. *Omega*, pp.155 – 164.

[18] May Gokan, et al. 2015. Multi-objective genetic algorithm for energy-efficient job shop scheduling. *International Journal of Production Research* **53(23)** pp.7071-7089.

[19] Duplakova D., Teliskova M., Duplak J., Torok J., Hatala M., Steranka J., &Radchenko S. 2018. Determination of optimal production process using scheduling and simulation software. *International Journal of Simulation Modelling,***17(4)**, pp.609-622.

[20] Coman A P D L., &Burian M. 2013. Minimizing make span in job shop production using a network approach. *Robotica& Management,***18(2)**. pp.234-246.

[21] Dagne T B., Jayaprkash J., &GeremewGebeyehu S. 2020. Design of Supply Chain Network Model for Perishable Products with Stochastic Demand: An Optimized Model. *Journal of Optimization in Industrial Engineering,***13(1)**, pp.29-37.

[22] Irman A., Febianti E., &Khasanah U. 2019. Minimizing makespan on flow shop scheduling using Campbel Dudek and Smith, particle swarm optimization, and proposed heuristic algorithm. *In IOP Conference Series: Materials Science and Engineering*, 673(1), p. 012099.

[23] Isa N A., Bareduan S A., &Zainudin A S. 2020. Bottleneck-Based Heuristic for Permutation Flowshop Scheduling. *In IOP Conference Series: Materials Science and Engineering,* 824(1), pp. 012-020

[24] Balo.g M., Duplakova D., Szilagyi E., Mindas M., &Knapcikova L. (2016). Optimization of time structures in manufacturing management by using scheduling software Lekin. *TEM Journal*, **5(3)**, pp.319-326

[25] Jayaprakash J., Thinakaran N., & Sathish T. 2014. JADE implementation of Multi-agent-based Inventory routing problem. *International Journal of Applied Engineering Research,* **9(24)**, pp.28785-28792.

IOP Conf. Series: Materials Science and Engineering **992** (2020) 012011 doi:10.1088/1757-899X/992/1/012011

Implementation of design for manufacturing in hot forging

P.Ruban and J.Jayaprakash

Department of Mechanical Engineering, Saveetha School of Engineering, Saveetha Institute of Medical and Technical Sciences, Chennai, India.

E-mail: agastinruban@gmail.com

Abstract. The manufacturing of components in Hot forging method is seen in detail. The concept of Design for Manufacturing in Product design and development is discussed. The Role of Design for Manufacturing will be implemented in this project for the reduction of cost of production, Reduce the wastage during production and also to increase the production rate. Here in this project the Six Case study will be taken for the design of manufacturing in Hot Forging. The cost evaluation in the hot forging process is evaluated.

Key words. Design for Manufacturing; Hot Forging; Product Design and Development.

1. Introduction

Design for manufacturing is a concept of Engineering which is widely used for product designing which is done for improving the easy fabrication of the product. This concept is almost followed in all the engineering disciplines, but implementation of concept differs widely depends on the manufacturing technology. Here in this project the design of manufacturing is been practiced in the Hot Forging Technology.

Forging is a type of manufacturing process where the metal is shaped as per the shapes in the Die. There are three type of Forging: Cold Forging, Warm Forging & Hot Forging. This manufacturing process of metal shaping can be done through Hammering, Pressing & Rolling. The General concept of forging is that the metal is deformed plastically to the desired Shape by giving a high fatigue resistance and strength.

1.1. Manufacturing Process:

The Manufacturing process of Hot Forging has the process flow of Billet Shearing, Billet Heating, Forging and Trimming & Shot blasting.

The Forging is formed in three stages such as Rolling, Moulding & Finishing. The stages of forging process are as follows:

Table 1 – Manufacturing processes of Forging

Process Name	Pictorial Representation
Rolling	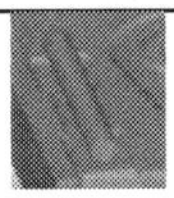 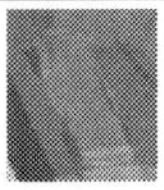
Moulding	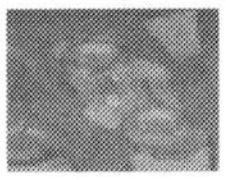
Finishing	

2. Literature Survey

The literature survey to be done for forging is broadly divided into various categories which are depending upon the research done as below: Various Techniques of Hot Forging, The modifications done in the Techniques, Results observed and the case studies have also been discussed.

2.1. Defect Removal:

The Major area of forging research carried out is the Defect removal technique because the greater number of losses occurs due to the rejections or rework of product to remove the forging defects in the forging industry [1]. The various types of forging defects that occurs on the Forged Product are lapping, mismatch, scales, Surface cracks, unfilling, etc. [5] These defects can be resulted due to the poor design or poor extrusion of manufacturing or also due to the material related problems. Those defects were been investigated and rectified by various researchers by doing a case studies on integral axle arm [7], Axially symmetrical and flanged components, Synchronizer ring, Stud bolt, Steel End Plate, etc.

2.2. Ease of Manufacturing:

The Capability of manufacturing a components is also an important aspect which ensures a greater importance in the industries which aids in the manufacturing of a part. This also helps in manufacturing a product with minimum forging load. The product can be executed only through a proper die design, choosing of proper billet shape and size and optimized level of process conditions [2]. These requirements are been explained through various case studies and researches taken up by many researchers which includes Forging parts like Wear Specimen and Centre Guide, helical forging, Alloy wheels, etc.

2.3. Process Optimization:

The process of Forging can be optimized by altering various process parameters like Die Stroke Speed, friction coefficient, temperature, etc. [6]. An optimized process is carried out to achieve a good quality of product with minimal forging loads [8]. Various components have also been studied like starter motor ring gears, compressor blades, helical gears, pneumatic clamp, spindles, etc.

IOP Conf. Series: Materials Science and Engineering **992** (2020) 012011 doi:10.1088/1757-899X/992/1/012011

2.4. Preform Design:

Preform design includes the factors like changing the billet shape, change of flash thickness and width, increasing the corner and fillet radii to reduce the forging loads and also to ensure the complete filling of Die impression [3]. A good preform design results in the proper distribution of metal in the die cavity. Many case studies have been taken up by many researchers include Pinion and Helical gear, Connecting rod, etc.

2.5. Die Analysis:

Die life plays an important role in the forging industry because the major cost of production is the die Cost [9]. Fatigue Analysis of the dies are done using the FEA which gives a good estimation of forging dies and tools life [4]. The surface of die is also an essential criteria because it improves the coefficient of friction which is involved in the metal flow during forging process.

3. Implementation Of Design For Manufacturing In Hot Forging

3.1. Forging Route Change

Material Used: The material used in this part is S10C.
Customer: PRF > SEAHAN STAMPING > HANON > GENEREAL MOTORS
Application: A Fan Pulley is mounted in car of Air Conditioner Sub-subheadings.

3.1.1. Problems Identified

- High Draft Angle.
- More Forged Weight compared to Machined Weight.
- Thickness not under control.

3.1.2. Recommendation:

- The Draft angle can be reduced.
- Thickness tolerance can be reduced.
- Mass of the part can be reduced.

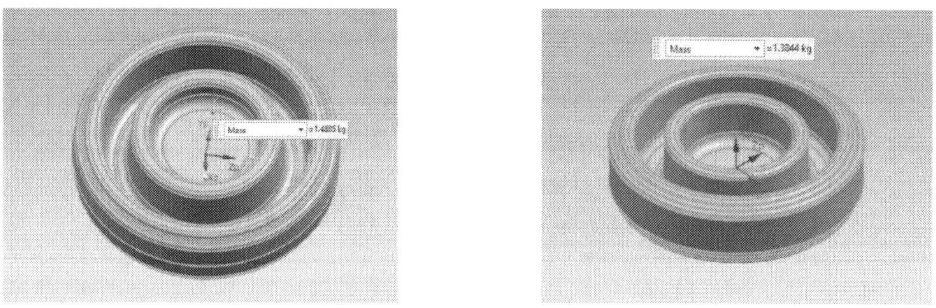

Figure 1: Pulley Forging
a) Hammer Route b) Press Route

3.2. Implementation of Forging Design

Material Used: The material used in this part is EN-8D.
Customer: PRF > RANE > TRACTORS AND FARM EQUIPMENT LIMITED
Application: The movement of Drop Arm results in motion of the stub axles.

3.2.1. Problems Identified:

- No Draft Angle provided
- No Ejection angle in opposite directions.
- The part is not angles vertically aligned.
- Sharp edges and No Radius provided in corners.

3.2.2. Recommendation:

- The Ejection angle has been provided.
- Orient profile has been provided.
- The Part is aligned vertically
- Sharp edges has been removed and radius has been provided.
  This is a preparatory step in which billet (rod) is compressed on one side for the amount of 32mm (Fig.2). This operation is foreseen for the mass distribution along the longitudinal axis of the work piece in order to enable better die filling in successive, following operation.

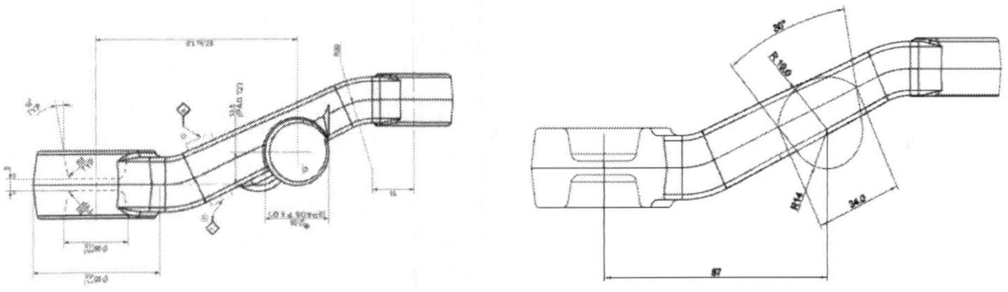

Figure 2: Design of Manufacturing in Forging
a.) Before Implementation of DOM b.) After Implementation of DOM

3.3. Commonization of Similar Parts:

Material Used: The material used in this part is EN-8D.
Customer: PRF > RANE > TATA MOTORS
Application: A roller follower having two or three teeth engages with the worm teeth and is carried on two sets of needle rollers supported on a short steel pin.

3.3.1. Problems Identified:

- The two parts has the similar profile and two die has been carried out.
- Two Setting time found for the two part forging process.
- The Mechanical properties and supply condition of the both parts found same.

IOP Conf. Series: Materials Science and Engineering **992** (2020) 012011 doi:10.1088/1757-899X/992/1/012011

3.3.2. Recommendation:

- The Two parts can be communized to a single part.
- The single die can be used for the production of both the parts.

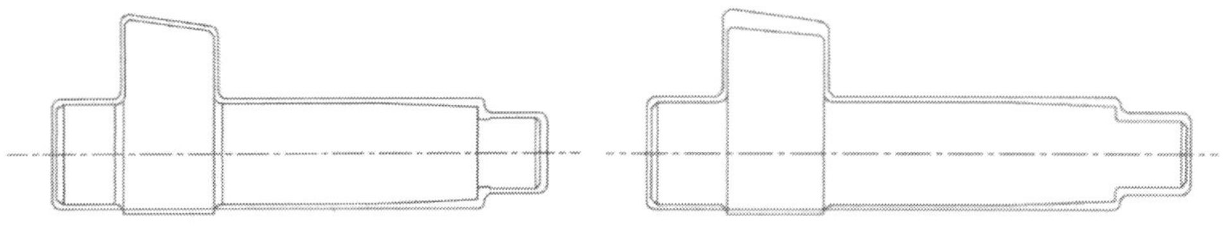

Figure 3: Superimposing of two machined 2D with Communized forging 2D
a.) Rocket Shaft 1 2D Super Imposed b.) Rocket Shaft 2 2D Super Imposed

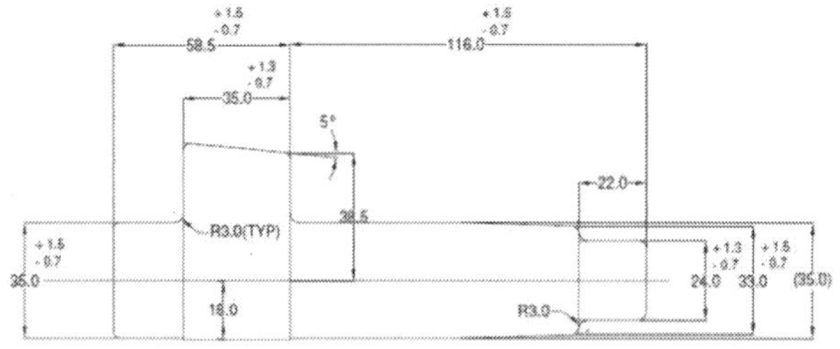

Figure 4: Final Forging drawing after Commonization

3.4. Common Moulder for LH & RH Finisher

Material Used: The material used in this part is EN-8D.
Customer: PRF > RANE > POLARIS
Application: The Tie rod end is used in the alignment of wheels which provides the adjustment for the wheel to keep the tires free from wear out on the either edges.

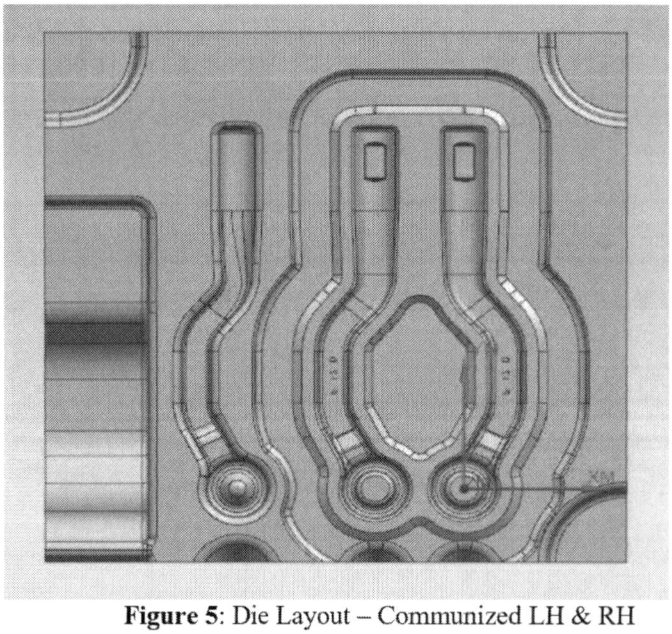

Figure 5: Die Layout – Communized LH & RH

3.4.1. Problems Identified:
- The two parts has similar profile with two orientation LH / RH and two die has been used.
- Two Setting time found for the two part forging process.
- The Mechanical properties and supply condition of the both parts found same.

3.4.2. Recommendation:
- The process stages of two parts can be communized in a single die.
- The common moulder can be used for the production of the both finisher.

3.5. Common Moulder for Long and Short Parts:

Material Used: The material used in this part is EN-8D.
Customer: PRF > RANE > NEXTEER > FORD MOTOR CORPORATION

3.5.1. Problems Identified:
- The two parts has the similar profile with only change of Length and two die has been used for the production.
- Two Setting time found for the two part forging process.
- The Mechanical properties and supply condition of the both parts found same.

3.5.2. Recommendation:
- The process stages of two parts can be communized in a single die.
- The common moulder can be used for the production of the both finisher.

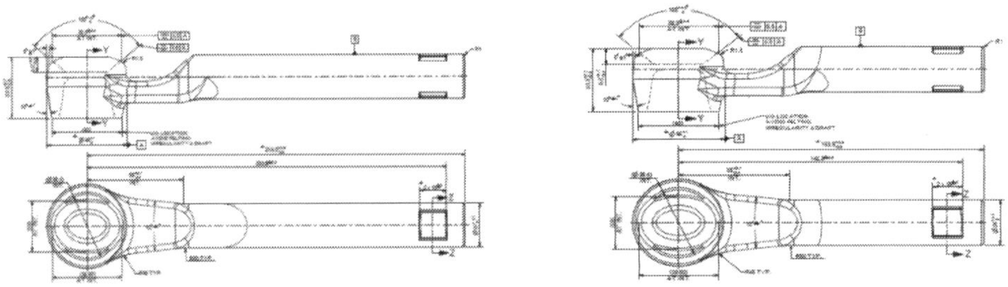

Figure 5: 2D Sketch of Socket Forging
a.) Socket Forging Long b.) Socket Forging Short

Figure 6: Communized Die for Long & Short Socket Forging

3.6. Two Impression in a Single Die

Material Used: The material used in this part is S35C.
Customer: PRF > RANE NSK > TATA MOTORS LIMITED
Application: Steering Yoke is the most common type of steering that is driven by the steering wheel from the user.

3.6.1. Problems Identified:

- The Smaller part covers only the 50% of the Standard Die size
- The Life of the parts gets completed in single die sinking is very less.
- The Excess Die area found unused.

3.6.2. Recommendation:

- The two set of all three stages is provided in the single die.
- Two rectangular lock is provided at two diagonal ends to control mismatch.

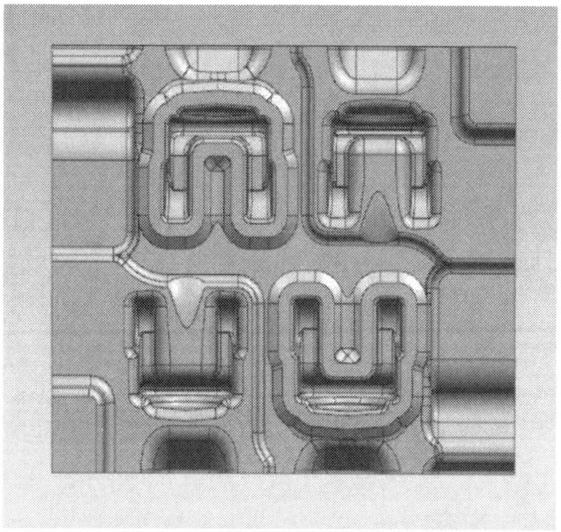

Figure 7– Die Layout with two set of Die Cavity

4. Conclusion :

Design for manufacturing is an important aspects of reducing overall costs during product development. The Role of Design for Manufacturing in this project has given the great reduction of cost of production, Reduce the wastage during production and also to increase the production rate [10].

Some of the important findings from the each case study is given below:

- In the first case study, we recommended to change the Forging process from Friction drop hammer forging to Mechanical press Forging. Due to this process modification, raw material used has been reduced up to 7% from the existing method and with the result of above change Cost reduction of Rs.5.54 / Part was achieved.
- In the second case study, we recommended a new design change in customer Part drawing based on the complexity in the forging process. This made the part feasible for production in hot

IOP Conf. Series: Materials Science and Engineering **992** (2020) 012011 doi:10.1088/1757-899X/992/1/012011

forging.

- In the third case study, we recommended different part drawings into common single die set which gave the cost reduction of 50% from the existing die cost.
- In the fourth case study, we recommended a single die set for two Left and Right Oriented part (Socket Forging) to reduce the cost of die and the die setting time.
- In the fifth case study, we recommended a single die set for different dimensional with similar part profile (Socket Forging) to reduce the cost of die and the die setting time.
- In the Sixth case study, we recommended the improvement in existing die layout and thereby doubled the production rate.

References

[1] Peter Christiansen, Jesper H.Hattel, Niels Bay and Paulo A.F.Martins, "*Modelling of Damage during Hot Forging of Ingots*", the 5th International Conference, Czech Republic (2013).

[2] P.S.Sambhunath and M.T.Sijo, "*Process Parameters Designing and Simulation for the Non-Isothermal Forging of Ti-6Al-4V Alloy*", International Journal of Emerging Technology and Advanced Engineering, Vol. **3**, Issue 8, ISSN 2250-2459, pp.331-338 (2013).

[3] Olivier Pantalé and Babacar Gueye, "*Influence of the Constitutive Flow Law in FEM Simulation of the Radial Forging Process*", Journal of Engineering, pp.1-8 (2013).

[4] Ming-Chao Sun, Gow-Yi Tzou and Liang-An Zheng, "*Processing Animation Simulation and FEM Analysis of Multi-stage Cold Forging of Stainless Automotive Battery Fastener*", Indian Journal of Engineering and Material Sciences, Vol. **20**, pp.219-224 (2013).

[5] Aju Pius Thottungal and M.T.Sijo, "*Controlling Measures to Reduce Rejection Rate due to Forging Defects*", International Journal of Scientific and Research Publications, Vol. **3**, Issue 3, ISSN 2250-3153, pp.1-6 (2013).

[6] Xinghui Han and Lin Hua, "*3D FE Modeling Simulation for Wear in Cold Rotary Forging of 20CrMnTi Alloy*", Journal of Tribology, Vol. **135**, pp.011101/1-15 (2013).

[7] Christy Mathew, Justin Koshy and Dr. Deviprasad Varma, "*Study of Forging Defects in Integral Axle Arms*", International Journal of Engineering and Innovative Technology, Vol. **2**, Issue 7, ISSN 2277-3754, pp.322–326 (2013).

[8] Iaroslav G.Zhbankov and Alexander V.Perig, "*Forging of Ingots without Hot Tops*", Materials and Manufacturing Processes, ISSN 1042-6914, pp.229-235 (2013).

[9] A.Gontarz, Z.Pater and K.Drozdowski, "*Hammer Forging Process of Lever Drop Forging from AZ31 Magnesium Alloy*", METALURGIJA, Vol. **52**, ISSN 0543-5846, pp.359-362 (2013).

[10] Sunil Mangshetty and Santosh Balgar, "*Billet Shape Optimization for Minimum Forging Load using FEM Analysis*", International Journal of Engineering Research and Development, Vol.**3**, Issue 3, ISSN 2278-800X, pp.11-16 (2012).

IOP Conf. Series: Materials Science and Engineering **992** (2020) 012012 doi:10.1088/1757-899X/992/1/012012

Design of reduced dipole yagi – Uda antenna

P. Muthu Kannan[1], M. Reji[2], R. Swaminathan[3]

[1] Professor, Saveetha School of Engineering, Saveetha Institute of Medical and Technical Sciences, Chennai, India.
[2]Associate Professor, Saveetha School of Engineering, Saveetha Institute of Medical and Technical Sciences, Chennai, India.
[3]Associate Professor, Saveetha School of Engineering, Saveetha Institute of Medical and Technical Sciences, Chennai, India.

E-mail: muthukannan@saveetha.com

Abstract. This work provides design and analysis of a dipole antenna whose height is reduced. Any Dipole antenna can be designed with $\lambda/2$ height. A half wave dipole is formed with '$\lambda/2$' height to get desirable characteristics of dipole antenna. Minimizing antenna height is always a desirable in antenna design. Hence, an attempt is made in this work to reduce the height of a dipole and as an application, Yagi-Uda antenna is constructed using this reduced dipole and optimized for better gain. The results are discussed in detail.

1. Introduction

The entire world became wireless over the last few decades. Huge technological developments are reported in the field of wireless communication. The communication systems are developed with extremely high SNR, Better throughput and vastly improved encryption techniques. Improvements in the design of antenna technology is really serving a lot in the implementation of effective wireless communication systems.

Varieties of antennas are developed such as Wire antennas, Microstrip antennas and Dielectric Resonator antennas [1]. Wire antennas are the basic antenna types. Lot of variants of Microstrip antennas such as fractal antennas, Reconfigurable antennas, Electromagnetic Bandgap antenna, Wearable antennas are discussed by Balanis [1], Kraus [2]. K.D. Prasad [3] and D.M. Pozar [4] [5].

A Dipole antenna is the basic antenna in the development of wire antennas. The height of dipole antenna is the design criterion. The height of the dipole antenna is given as "Half of the wavelength" at the resonant frequency. In the design of microstrip based dipole antenna also the above design is used. If this height of the antenna can be reduced while maintaining the same desirable characteristics, it will be useful in saving the required space, which in turn will produce miniaturized antenna.

The discussion of Microstrip dipole antenna on various ground planed is discussed [6] and design of Microstrip dipole antenna for WLAN is elaborately explained [7].With the design of reduced dipole, it is possible to design a Yagi – Uda Antenna which is a popularly used antenna for Television reception. The reduction in size of Yagi-Uda antenna by modifying the diameter and spacing of the elements is discussed [8]. Sheldon and Bray pointed out printed Quasi- Yagi antennas with reduced mass [9]. D.I. Savitri compared hollow aluminium Yagi- Uda antenna and Dipole antenna for Indoor Television reception [10].

IOP Conf. Series: Materials Science and Engineering **992** (2020) 012012 doi:10.1088/1757-899X/992/1/012012

2. The Problem

The design parameter of a dipole is the height of the antenna which is calculated as half of the wavelength at resonant frequency. For example, a dipole antenna resonating at 300 MHz, should have a height of 0.5 m which is a sizeable dimension. The required space for the development of antenna is also depending on this height.

If it is possible to reduce the height of the antenna, with the same characteristics, the required space for the development of antenna can be reduced. This will help in the development of miniaturized antenna. This work was carried out to design a dipole antenna at 300 MHz with reduced height and a Yagi – Uda antenna is also designed and analyzed using this reduced dipole.

3. Design of dipole antenna

To design a dipole antenna, a freeware MMANA-GAL is used which is a reliable software for the design of wire antennas. The dipole antenna is selected to resonate at 300 MHz. The height of the antenna is calculated as the half of the wavelength. For a 300 MHz frequency, the corresponding wavelength will be 1 m and half wavelength will be 0.5 m. The antenna is resonating in free space condition and is assumed to be made up of loss less material. By employing the tuning mechanisms available in the software, the height of the designed antenna is given as 0.476 m and oriented in 'z' axis at 0.238 m on either side. The designed antenna is shown in the figure 1.

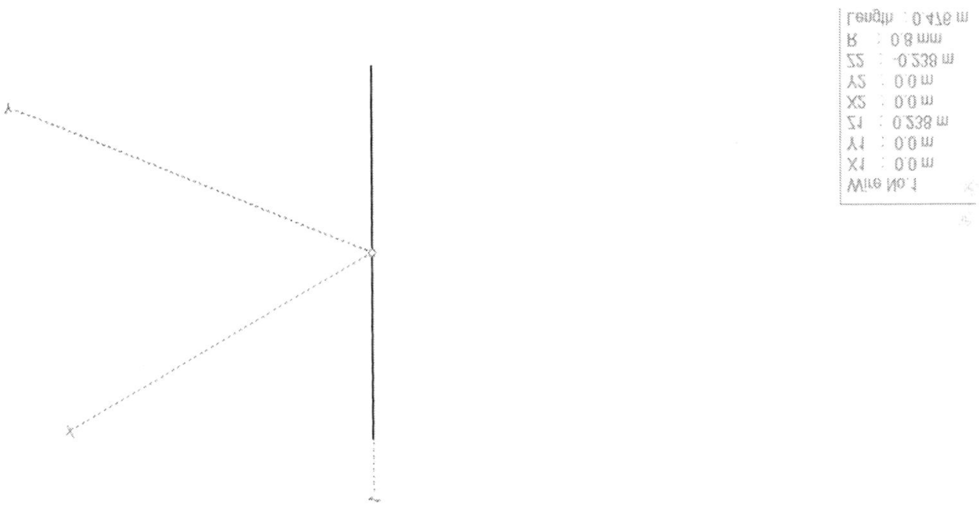

Figure 1. Dipole antenna at 300 MHz

4. Dipole antenna and reduced dipole antenna

4.1 Design of Dipole Antenna

The tuned dipole is given in Figure 1 which resonates at 300 MHz. The radiation pattern and VSWR are calculated using the software mentioned.

IOP Conf. Series: Materials Science and Engineering **992** (2020) 012012 doi:10.1088/1757-899X/992/1/012012

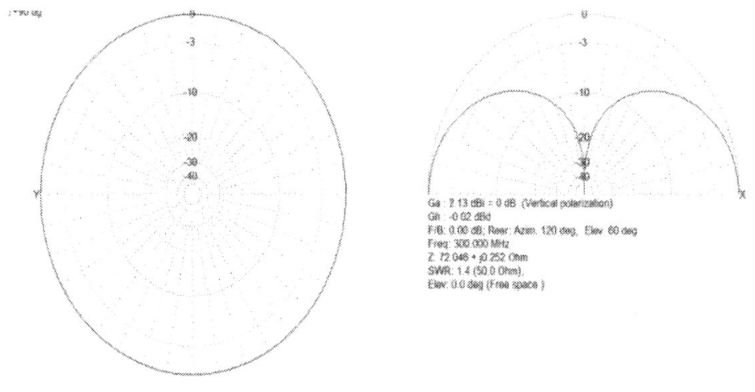

Figure 2. Radiation pattern of normal dipole antenna

As per the results shown in Figure 2, this dipole antenna has impedance of 72 + j 0.252 Ω which is close to the impedance of a dipole which is 73 Ω. Hence this antenna resonates well which is also evident from the SWR given as 1.4. The gain is given as 2.13 dBi.

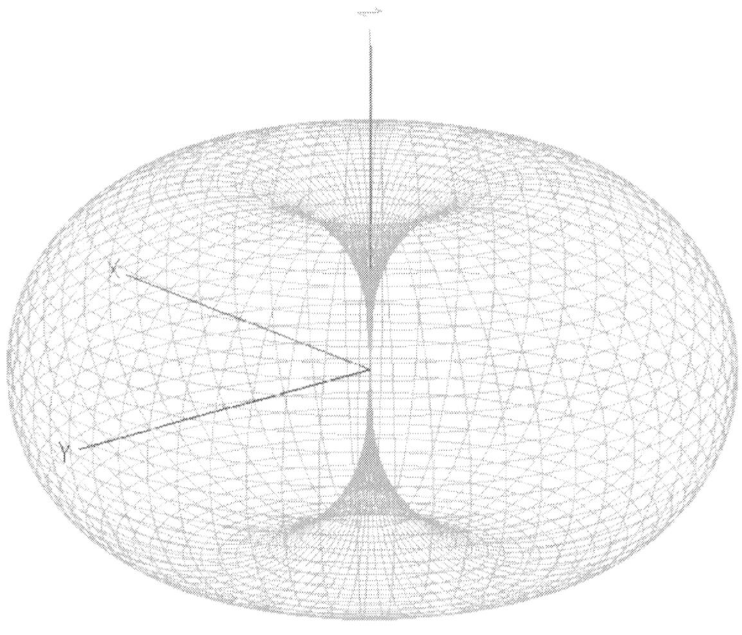

Figure 3. 3D Radiation pattern of normal dipole antenna

Figure 3 gives the three dimensional radiation pattern of the dipole antenna at 300 MHz. The radiation patten obtained is a proper three dimensional pattern.

4.2 Design of Reduced Dipole Antenna

The height of the antenna is to be reduced but with similar characteristics. The antenna which is designed has a height of 0.396 m in the 'Z' direction which is shown in Figure 4. This antenna is designed with four wires with the edge of the dipoles are placed at 0.193 m on either side which gives the overall height of 0.396 m.

IOP Conf. Series: Materials Science and Engineering **992** (2020) 012012 doi:10.1088/1757-899X/992/1/012012

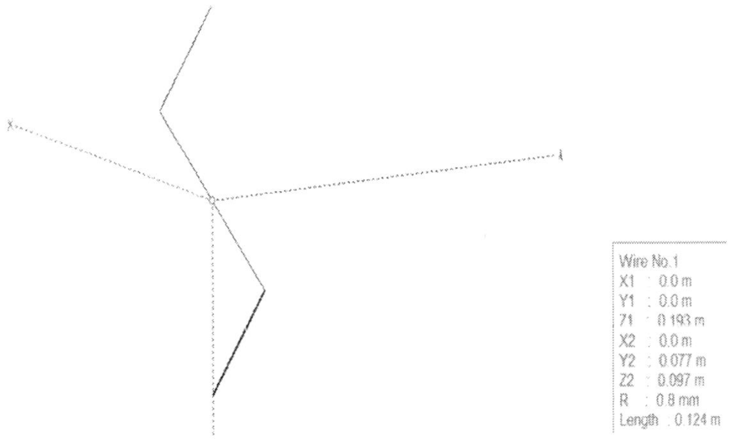

Figure 4. Reduced dipole antenna at 300 MHz

The antenna given in Figure 4 is made up of four wire segments. Each wire segment has a length of 0.124 m which gives the overall length of the antenna as 0.496 m. Even though the length of 0.5 m for a dipole antenna is maintained, the height of the antenna is reduced to 0.08 m.

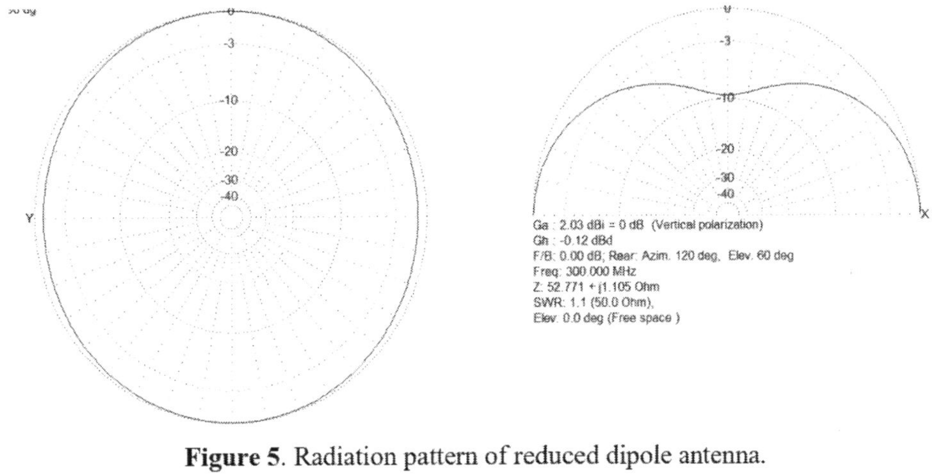

Figure 5. Radiation pattern of reduced dipole antenna.

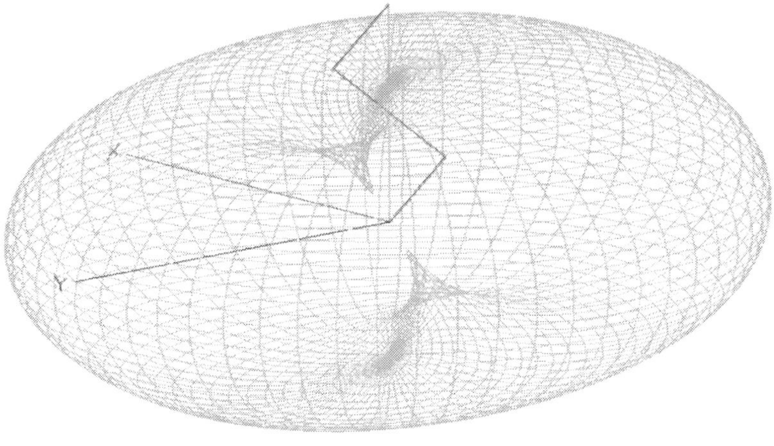

Figure 6. 3D Radiation pattern of reduced dipole antenna.

From the figure 5, the gain of the antenna is given as 2.03 dBi and the input impedance is given as 52.7 + j 1 Ω. The SWR is given as 1.1. These characteristics give a good similar result. The three dimensional radiation pattern of the above dipole is given in Figure 6.

5. Design of Yagi- Uda antenna with normal and reduced dipoles

5.1 Design of Yagi-Uda Antenna with normal dipole

As an application with the reduced dipole, design of yagi antenna is considered in this work. Yagi antenna is a directional antenna which is widely used in Television reception in RF range. A yagi antenna has a dipole or folded dipole as an active element with the length of 0.5 m. It has a director whose length is five percent shorter than that of active element and a reflector whose length is five percent longer that of active element. The distance between director and active element is modified as 0.2 m and the distance between active element and the reflector is also fixed as 0.2 m.

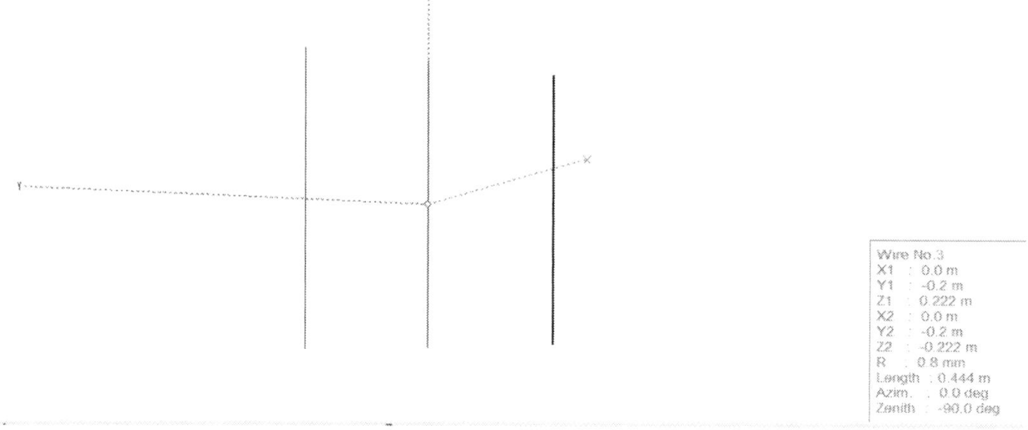

Figure 7. Yagi antenna with normal dipole

The two dimensional radiation pattern of the Yagi antenna at 300 MHz with normal dipole is given in Figure 8 and three dimensional radiation pattern is given in Figure 9.

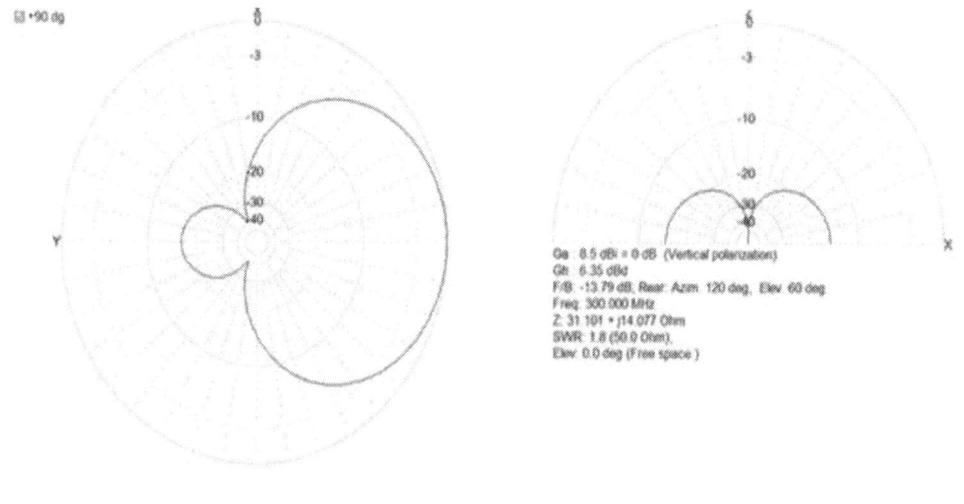

Figure 8. Radiation pattern of Yagi antenna with normal dipole at 300 MHz

IOP Conf. Series: Materials Science and Engineering **992** (2020) 012012 doi:10.1088/1757-899X/992/1/012012

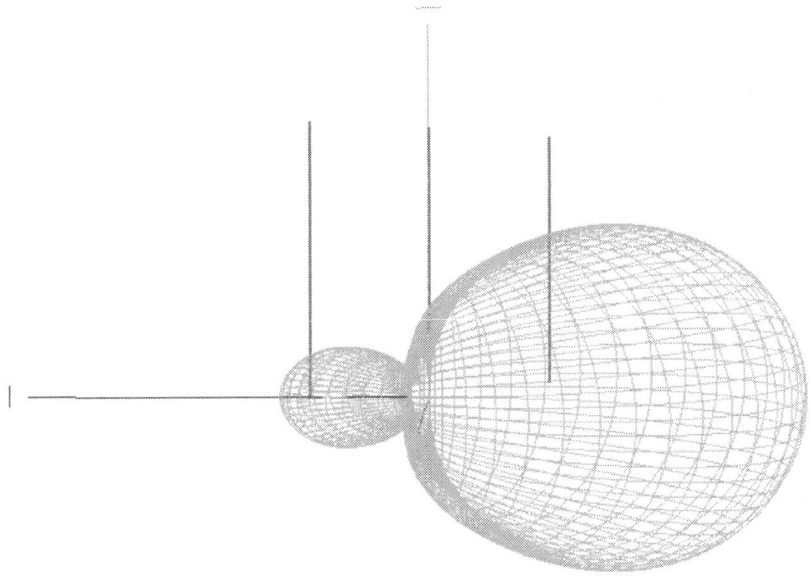

Figure 9. 3D Radiation pattern of Yagi antenna with normal dipole at 300 MHz.

5.2 Design of Yagi- Uda Antenna with reduced dipole

Now, design of Yagi antenna is carried out with reduced dipole and the reflector and directors are having the same length and spacing as that of Yagi antenna with normal dipole. The structure is shown in Figure 10 at which the active element is the same as shown in Figure 4.

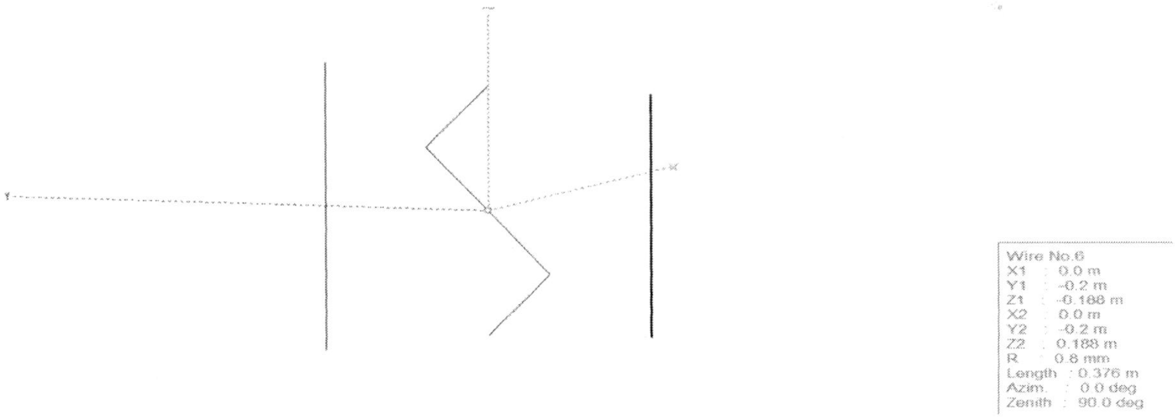

Figure 10. Yagi antenna with reduced dipole

Radiation pattern of Yagi antenna with reduced dipole is shown in Figure 11. The corresponding three dimensional pattern is also given in Figure 12.

IOP Conf. Series: Materials Science and Engineering **992** (2020) 012012 doi:10.1088/1757-899X/992/1/012012

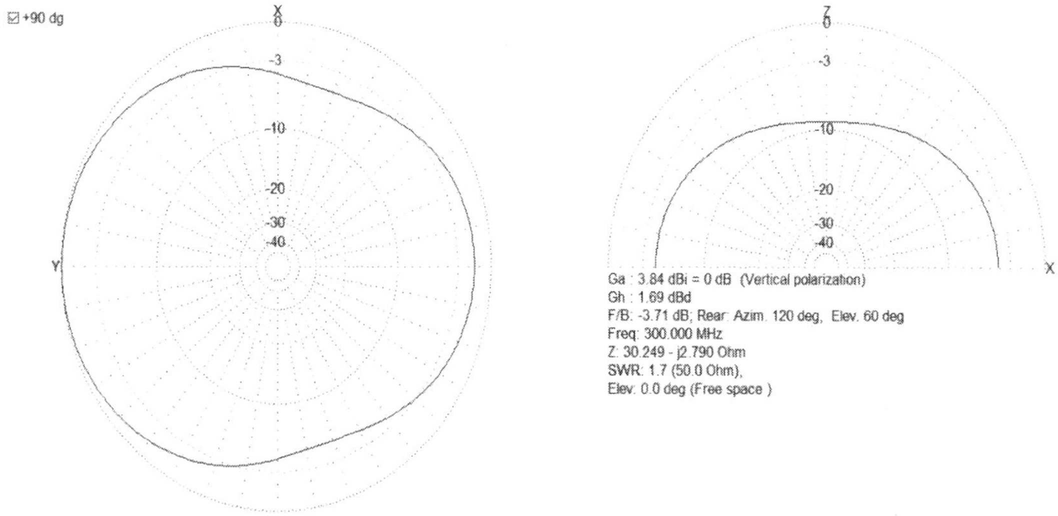

Figure 11. Radiation pattern of Yagi antenna with reduced dipole.

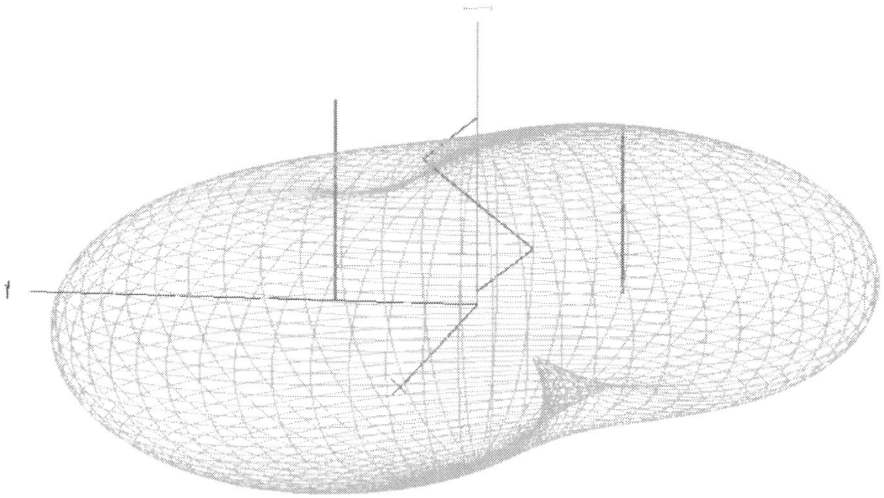

Figure 12. 3D Radiation pattern of Yagi antenna with reduced dipole

Comparing the radiation pattern and directivity of Yagi antenna with normal dipole and reduced dipole, the gain (or) directivity is reduced in reduced dipole for the same dimensions. It is important to improve the gain as much as possible. Hence it is decided to reduce the distance between the active element and reflector. But the gain is reduced even further. Then, the distance between the active element and reflector is increased. The gain is increased compared with Yagi antenna with reduced dipole as given in Figure 11 and 12.

First, the Yagi antenna with reduced dipole is maintaining other dimensions but the spacing between active element and the reflector is increased from 0.2 m to 0.3 m; The structure of the antenna is given in figure 13.

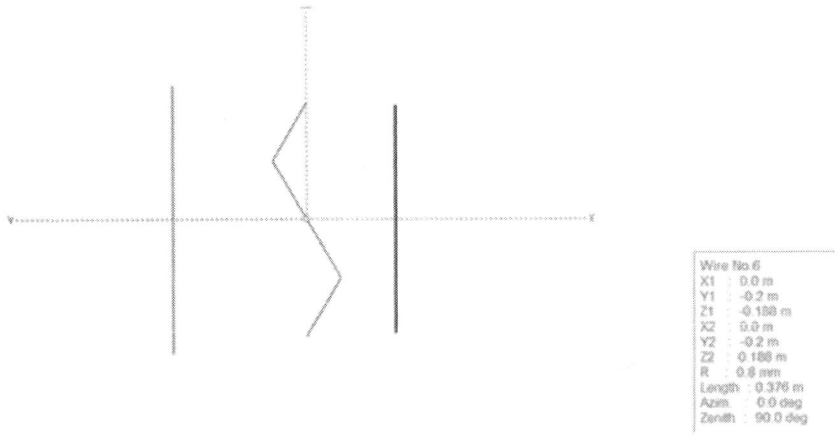

Figure 13. Yagi antenna with reduced dipole and Reflector spacing of 0.3 m.

The radiation pattern of Yagi antenna with reduced dipole for a reflector spacing of 0.3 m is shown in Figure 14. It gives marginal improvement compared to Figure 11. The gain is increased to 4.97 dBi from 3.84 dBi given in Figure 11.

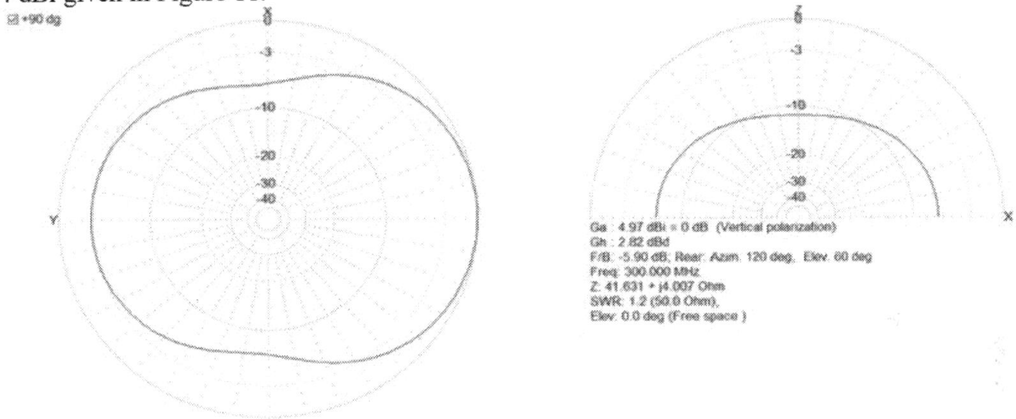

Figure 14. Radiation pattern of Yagi antenna with reduced dipole and reflector spacing as 0.3 m

Hence it is decided to further increase the distance between the active element and the reflector and the same is increased to 0.4. The structure of the Yagi antenna is given as shown in Figure 15.

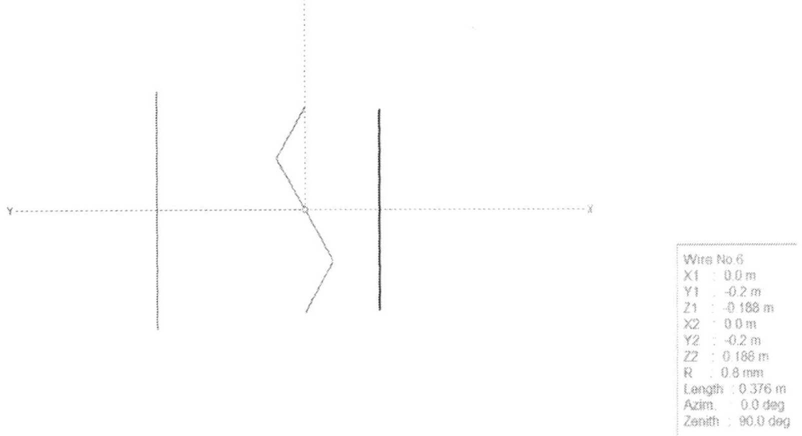

Figure 15. Yagi antenna with reduced dipole and reflector spacing of 0.4 m.

The radiation pattern of the Yagi antenna with reduced dipole and reflector spacing of 0.4 m is given in the Figure 16. The gain is marginally reduced to 4.9 dBi.

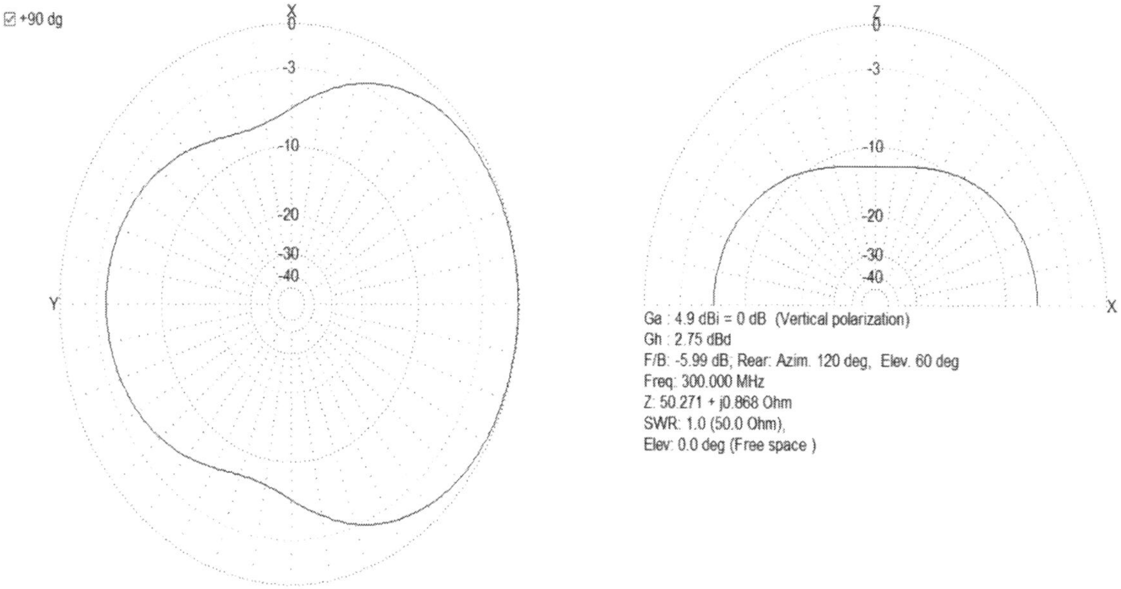

Figure 16. Radiation pattern of Yagi antenna with reduced dipole and reflector spacing 0.4 m.

Based on the reflector spacing in the previous two cases, it is observed that the spacing between active element and reflector of 0.3 m, has better gain of 4.97 dBi. But this gain is less compared with Yagi antenna with normal dipole which is 8.5 dBi. Hence there are few methods to be tried to increase the gain of Yagi antenna. For a reflector spacing of 0.3m for which the gain is maximum, the length of reflector is increased to 0.5 m and the radiation pattern is found out. The gain is given as 5.26 dBi as shown in Figure 17.

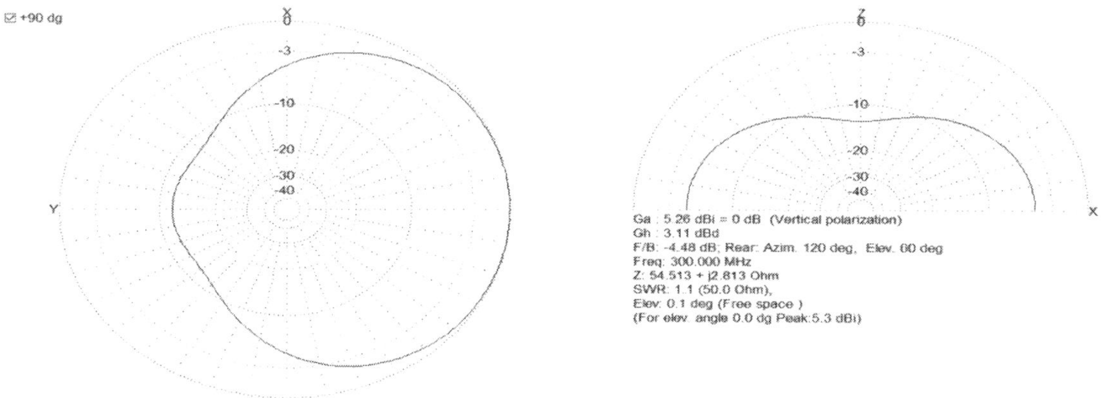

Figure 17. Radiation pattern of Yagi antenna with reduced dipole and reflector length of 0.5 m.

The reflector length can be varied further to find out the maximum gain from Yagi antenna with reduced dipole. The reflector length is increased to 0.54 m and the gain achieved is 4.34 dBi. When the reflector length is increased to 0.6 m, the gain is reduced to 3.84 dBi. Hence, maximum gain is achieved with the reflector length of 0.5 m and the active element – reflector spacing of 0.3 m. But keeping the active element - reflector spacing of 0.2 m and reflector length of 0.5 m, gives even better gain of 5.94 dBi. The radiation pattern in shown in Figure 18. Hence the optimum spacing of active element – reflector is considered as 0.2 m for the reflector length of 0.5 m.

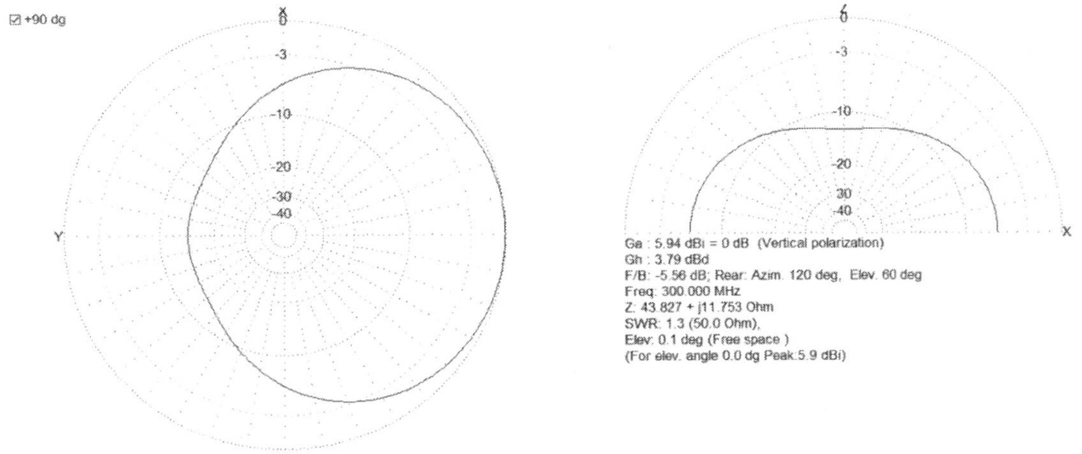

Figure 18. Radiation pattern of Yagi Antenna for reflector length 0.5 m and spacing 0.2 m

6. Conclusion

The design of a dipole with reduced height is designed and discussed. The height of the dipole is reduced with 0.08 m. But almost same gain is obtained with good standing wave ratio. This design will help in miniaturized dipole antenna. Even though the design was tested in free space condition as wire antenna, the same design may be extended to microstrip antennas also. Hence it is possible to reduce the height of the antenna and in turn the space required for conventional dipole antenna. Using this antenna, an application of Yagi-Uda antenna is considered. The radiation pattern and gain are found out for various conditions. It is found out that reflector length of 0.5 m and active element – reflector spacing of 0.2 m gives maximum gain of 5.94 dBi which is considered as an optimized solution for this reduced dipole.

References

[1] C.A.Balanis, "Antenna Theory: Analysis and Design", 3rd edition, John Wiley, 2009.

[2] J.D.Krauss, "Antennas", 2nd edition, Tata Mcgraw Hill, 1997.

[3] K.D. Prasad, Deepak Handa, "Antenna and Wave Propagation", 3rd edition, Satyaprakashan, 2003.

[4] D.M.Pozar, " Microwave Engineering", 3rd edition, John Wiley, 2005.

[5] D.M. Pozar, "Microstrip Antennas", Proceedings of IEEE, Vol. 80, No.1, pp 79-91, 1992.

[6] N. Karg, Z. aijaz, "Design of Microstrip dipole antenna at various ground plane", International Journal of Innovative research in Science, Engineering and Technology, Vol. 1, Issue 2, 2012.

[7] M.H. Jamaluddin et al., "Microstrip Dipole antenna for WLAN applications", IEEE, 2005.

[8] S. Daya Murali et al., "Size reduction of Yagi-Uda Antenna by altering the diameter and spacing between the elements", International Journal of Applied Engineering Research, 9(22), pp. 12011

– 12022.

[9] C.Sheldon, M. Bray, "Reduced mass printed Quasi – Yagi Antennas", 2016 IEEE International Symposium on Antennas and Propagation, October 2016.

[10] D.I. Savitri, " Comparison between Hollow Aluminium Yagi – Uda and Dipole Antenna for Indoor Antenna TV". IOP Conference series: Materials science and Engineering, 453, 2018.

Influence of nano SiO_2 in cement mortar using taguchi analysis

P Vasanthi[1] and S Senthil Selvan[2]

[1]Research Scholar, Sathyabama Institute of Science and Technology, Department of Civil Engineering, Chennai- 600 119, Tamilnadu, India.

[2]Department of Civil Engineering, SRM institute of Science and Technology, Chennai-603 203, Tamilnadu, India.

E-mail:vasanthihh@gmail.com

Abstract. In this research, optimization technique was employed to optimize the nano SiO_2 in cement mortar preparation by using Taguchi-technique. The main aim of this present investigation is to find the optimal input variables like nano SiO_2, w/c ratio and plasticizer to achieve cement mortar with high compressive strength and low water absorption percentage. When the input variables are more than one Taguchi - technique approach is a best method since it may give difference set of optimum level for each response. It is found that the addition of nano SiO_2, in cement mortar influenced high compressive strength with lower water absorption percentage. According to analysis nano-silica of 2wt. %, w/c ratio of 0.5 and plasticizer of 1wt. % (A2, B3 and C2) is found to be producing highest compression strength of 51.45MPa and lower water absorption percentage of 2.95.

1. Introduction

Cement is the most sustainable raw material in building industry with ever-increasing demand and performance [1]. New types of cement have been developed over the years as an alternative to conventional cements to increase the performance of conventional cement [2]. To enhance long term performance and adhesive properties the nano-particles could be added with cement in considerable volume [3]. This is because the growing demand for use in nano materials in recent years has brought growing popularity. Nano-oxides, ash fly, slag, and silica fume have been applied by many researchers to strengthen materials dependent on and without cement [4]. Ltifi et al. (2011) added the amorphous nano silica particle at a rate of 3 and 10% by mass of cement. According to author's conclusion the nano-SiO_2 improves the strength of cement pastes and develops the process of cement hydration. Moreover, the compression strength increases with increase of nano-SiO2 content [5]. Rahmani et al. (2012) found the abrasion strength increased during compression testing, the increase in water-cement ratio of micro silica concrete resulted in greater compressive strength [6]. Senff et al. [7] analysed the effect of silica fume (SF) and nano-silica (nS) on range of flow table, effective apparent porosity, and compressive strength and water absorption, un-restrained condition of shrinkage and weight loss of mortars up to 28 days using factorial design based experiments approach. Mortar with 7% of nano-silica exhibited quicker formation of structures in rheological test. The apparent porosity, compressive strength and water absorption are improved with the addition of nS 0-7%, 0-20% of SF and 0.35-0.59 of water/binder ratio. Similar study was done by Manzoor et al. [8]. The author has pointed out the use of RSM in optimizing concrete specimens by increasing compressive strength and decreasing RCPT and sorptivity. Liu et al. [9] calculated the effect of $CaCO_3$ nano particles (NC) with 2% and 1% fractional replacement of cement. It was verified that the rise in NC lowered the flow rate and shortened the specified time. Nonetheless, NC had no effect on standard cement quality water requirement. At the age of seven days and twenty-eight days, the bending strength and compression strength of toughened cement paste with NC is maximized and the optimum NC level was only 1%.

A review of nano technology in concrete was made by Sanchez et al. [3]. The author examined the advantages of nano technology in concrete technology and its applications in modern structural industries. Moreover the results indicated that nanotechnology can boost concrete quality and promote the improvement of new sustain and progressive cement composites with improved mechanical, thermal and electrical conductivity properties. Mosaberpanah et al. [10] attempted an experimental work with the use of nano silica the ultra-high performance concrete (UHPC) strength was increased via RSM optimization. The RSM results indicated that the accumulation of nano-silica improved the drying shrinkage and 28 days compressive strength and the reduction in the flow-ability.

While the flow-ability of fresh UHPC was reduced due to the addition of nano-silica. It is found that the addition of as received nano-silica particle in cement mortar forming agglomeration during settling process. This agglomeration could initiate crack development and reduce the compression strength of mortar [8]. To avoid agglomeration and ensure uniform settling of nano-silica particle, surface modification could be an effective and needful process [11]. In surface-treatment the local attraction between nano-silica is reduced via silane attachment. This silane attachment just covers the nano-silica as a coat and prevents clustering effect. This phenomenon may improve the performance of nano particle addition in base material [12, 13]. It is known that the addition of nano-silica could improve the compression strength of mortar however; high water absorption behavior of nano-silica could affect the life of cement mortar. In surface-modification process the nano-silica is converted as a hydrophobic one from hydrophilic nature. This phenomenon would influence good compression strength along with low water absorption in cement mortar. Hence, it could be the add-on advantage for modern infrastructure industries. Because making of high compressive strength cement mortar with low water absorption yield cost effective and economical process compatibility.

Taguchi optimization could be an effective tool when more number of input variables is employed in any engineering process. In this method different set of optimum level for each response could be identified at the end of optimization. The processing and mathematical modelling of Taguchi method is simple and effective on compare with other optimization techniques. Muthuramalingam et al. [14] have done an optimization study for micro machining process to optimize the process variables. According to author's conclusion the Taguchi optimization was the effective and fastest tool to find out the optimal value of inputs and finding of dominant variable. Similarly, Jayakrishnan et al. [15] optimized the end milling of aluminium alloy input parameters using Taguchi optimization technique. The authors concluded that the Taguchi optimization is the suitable method when multi responses are arrived from the particular process. Thus the previous literature revealed that there are lot of efforts have been made by many researchers for developing different types of micro and nano particle blended cement mortar. However, very few research works is presented based on surface modified nano addition along with optimization.

Based on previous literature there are lot of efforts have been made by many researchers for developing different types of micro and nano particle blended cement mortar. However, very few research works is presented based on Taguchi-technique optimization for optimizing the process variables in new type cement mortar. Based on the research gap the present study is designed. The chief aim of this research work is to examine and optimize the influence of input process parameters and to find optimum blending proportions of cement mortar with nano silica, water cement ratio (w/c) and plasticizer using Taguchi-technique analysis. Using design expert software (DES) the model built was checked for statistical adequacy and predictive reliability by means of variance analysis (ANOVA). These optimized process parameter cement mortar could be used as highly sustainable construction material for today's economical world [16]. Few research are followed in silane surface - treatment method [17, 18].

2. Experimental work

2.1. Materials

The ordinary port-land cement (OPC) was used. The OPC measures a specific gravity and fineness of 3.1 and 230 m^2/Kg. The cement paste used here was about 31% of standard consistency. A fine aggregate with a specific gravity of about 2.62 and fineness modulus of 3.98 was obtained for this study. Polycarboxylate was used as a super plasticizer (SP) and potable water was used for mixing nano silica of size 20nm with purity of 99.44%.

2.2. Cement Mortar and specimen making

The experimental work was carefully designed and to study the characteristics of nano SiO_2 on the cement mortar properties of different water cement ratios. Super plasticizer content was maintained constant for the entire range of mix. The nano SiO_2-mortar strength depends on the water cement ratio and the nano SiO_2 percentage on mix. The mix proportion adopted was C: FA=1:3 of all kinds. The samples were prepared as total content of cement was 200gms and sand 600gms, w/c ratio of 0.48, 0.50 and 0.52 and super-plasticizer content of 0.5 and 1wt. %. Polycarboxylate was solidified in water and the silica nano particles were added and stirred at a high speed for about 2 to 3 minutes. This mixture was added with cement mortar and filled with ASTM cube specimen (50mm ×50mm ×50 mm) using a vibrating table [19].

2.3. Testing of samples

The prepared cement mortar samples were tested for void content and compressive strength. The void content percentage was calculated using theoretical and measured density values according to ASTM C 1754. For void content test the w/c ratio of 0.5, plasticizer of 1wt. % and nano SiO_2 content of 0, 1.5 and 3wt. % was used as process parameters. However, the compressive strength of nano-silica toughened cement mortar is measured using a compressive strength tester according to ASTM C 109. Figure.1 shows the prepared cubes tested machine setup. Similarly, the water absorption percentage was calculated using ASTM C 1585. However, the compression strength and water absorption percentage of cement mortar was optimized using Taguchi technique design with 9 trials to find the optimum w/c ratio, plasticizer content and nano SiO_2 since these two parameters plays a major role in structure-property relationship. Table 1 shows the selection of process parameters and levels.

Figure 1. Cement mortar testing machine

Table 1. Process parameters and levels

Parameter	Symbol	Level 1	Level 2	Level 3
Nano SiO_2 (wt %)	A	1	2	3
W/C ratio (%)	B	0.4	0.45	0.5
Plasticizers (%)	C	0.5	1	1.5

3. Taguchi Method

Taguchi method was used to determine the optimal combination of input parameters in a systematic approach. In this approach L9 orthogonal array was used to find the optimal parameter. Since the 3input parameters and the L9 array was deployed to execute the optimization process. Based on his L9 totally nine different kinds of cement mortar was prepared and tested for compression strength and water absorption percentage. The compression strength was calculated using a compression tester whereas the water absorption was found using the weight of cubes before and after immersion in water. The quality analysis of the output responses are valued by signal to-noise ratio (S/N). This ratio is measured from the responses using suitable formulae. There are two categories of S/N ratios. They are smaller the better and larger the better. According to this present investigation the main aim is to improve the compression strength with minimal water absorption percentage. That means compressive strength is considering as larger the better whereas water absorption is consider as smaller the better. Equation 1 and 2 shows the S/N ratio of smaller the better and larger the better formulae.

For smaller the better

$$S/N\ ratio = -10 \log (1/n) + (Yij^2) \quad (1)$$

For larger the better

$$S/N\ ratio = -10 \log (1/n) + (1/Yij^2) \quad (2)$$

Where,

Yij – Values of response, n - Each experiment number of replication say '1'

Table 2. Design factor of L9 Orthogonal array and S/N ratio for Compressive strength and Water absorption

Experimen -tal Trail	Nano SiO_2 (wt%)	W/C ratio (%)	Plasticizers (%)	Compressive strength	Water absorpti- on	S/N ratio for CS	S/N ratio for WA
L1	1	0.4	0.5	35.80	5.29	31.08	-14.47
L2	1	0.45	1	36.86	5.11	31.33	-14.16
L3	1	0.5	1.5	37.05	5.05	31.38	-14.07
L4	2	0.4	1	45.90	3.27	33.24	-10.30
L5	2	0.45	1.5	43.22	3.98	32.71	-12.00
L6	2	0.5	0.5	49.66	3.02	33.92	-9.60
L7	3	0.4	1.5	41.55	3.92	32.37	-11.86
L8	3	0.45	0.5	42.57	3.80	32.58	-11.60
L9	3	0.5	1	50.21	3.45	34.02	-10.76

For this present study compression strength is larger the better and for water absorption smaller the better approach was used. This consideration is assumed since for a good cement mortar the compressive strength should be higher with minimal water absorption percentage. Table 2 depicts the L9 orthogonal array for response, the normalized S/N ratio and S/N ratio values calculated from the experimental response.

4. Results and discussion

This research work analyses and studies the process parameters effect on water absorption and compressive strength of cement mortar composites. The parameters are nano SiO_2 (A), w/c ratio (B), plasticizers (C) at three levels as shown in table 1.The degrees of freedom (DOF) for process factors were found to be 8 (DOF = 9−1 = 8).

The hardness and tensile strength was found using Taguchi method by using 16 investigational conditions [20].Table 3 and 4 depicts the response table for compressive strength and water absorption. The results showed that nano silica have correlation on cement mortar strength [16]. The better performance was observed for compressive strength and water absorption at higher S/N ratio as shown in Figure 2 and Figure 3. Due to the addition of surface-modified nano-silica particle of 2wt. % in cement mortar, which could fill in the voids of mortar and reduces the shrinkage and improves the compression strength [21]. The response value A2, B3, C2 at Optimum process condition are represented in table 2.

Table 3. Response table for S/N ratio: Larger is better (Compressive strength).

Level	Nano SiO_2 (wt %)	W/C ratio (%)	Plasticizers (%)
1	31.26	32.23	32.53
2	33.29	32.21	32.86
3	32.99	33.10	32.15
Delta	2.03	0.89	0.71
Rank	1	2	3

The optimized process parameters for compressive strength and water absorption are nano SiO_2 at level 2, w/c ratio at level 3, and the plasticizers at level 2.

Table 4. Response table for S/N ratio: Smaller is better (Water absorption).

Level	Nano SiO_2 (wt %)	W/C ratio (%)	Plasticizers (%)
1	-14.23	-12.21	-11.89
2	-10.63	-12.59	-11.74
3	-11.41	-11.47	-12.64
Delta	3.60	1.11	0.90
Rank	1	2	3

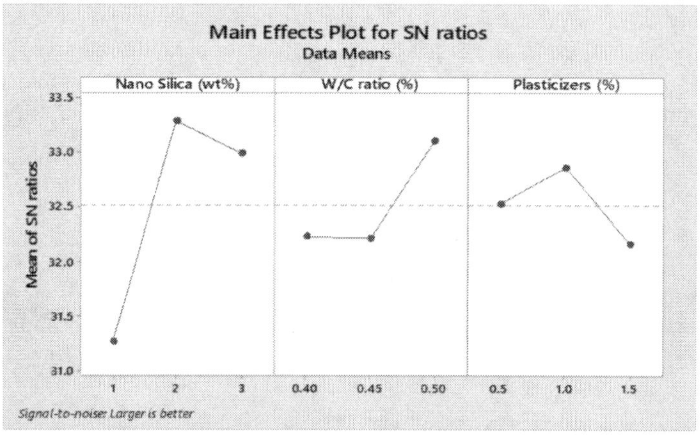

Figure 2. S/N ratio response curve of Compressive Strength.

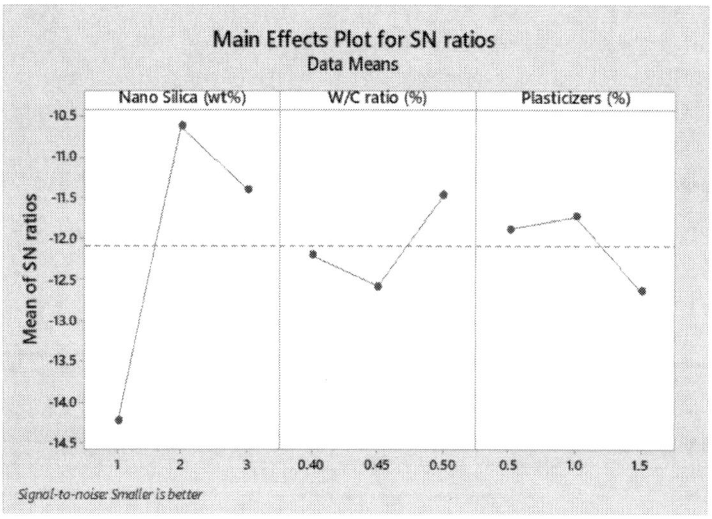

Figure 3. S/N ratio response curve of Water absorption

4.1. Analysis of variance

The experimental analysis and modelling results are obtained using the taguchi method that is a combination of arithmetical and computational techniques, which are used for analysis of results. The ANOVA table specifics 98.46% and 98.43% of significant level in compressive strength and water absorption behavior.

ANOVA table is determined by the total sum of squares of the standard deviation equal to the sum of squares of the standard deviation produced by the respective parameter [15]. Tables 5 and 6 represents the ANOVA results for compressive strength and water absorption respectively. It determines the F-ratio and P-value. The process parameter, the influence of nano SiO_2 plays a significant role in determining the mechanical behaviour. Compressive strength and water absorption improved by 20.33% and 21.92% respectively. Highest contribution of 88.76% was observed due to Nano SiO_2, W/C ratio and plasticizers were 18.86 and 9.05 % contributions.

Table 5. ANOVA variance table for Compressive Strength.

Source	DF	Seq SS	Contribution(%)	Adj SS	Adj MS	F-Value	P-Value
Nano Silica (wt%)	2	163.451	71.05	163.451	81.726	68.48	0.014
W/C ratio (%)	2	43.383	18.86	43.383	21.692	18.18	0.052
Plasticizers (%)	2	20.820	9.05	20.820	10.410	8.72	0.103
Error	2	2.387	1.04	2.387	1.193		
Total	8	230.041	100.00				

(R-Seq 98.96%)

Table 6. ANOVA variance table for Water absorption.

Source	DF	Seq SS	Contribution (%)	Adj SS	Adj MS	F-Value	P-Value
Nano Silica (wt%)	2	5.09332	88.76	5.09332	2.54666	56.36	0.017
W/C ratio (%)	2	0.32939	5.74	0.32939	0.16470	3.65	0.215
Plasticizers (%)	2	0.22524	3.93	0.22524	0.11262	2.49	0.286
Error	2	0.09036	1.57	0.09036	0.04518		
Total	8	5.73831	100.00				

(R-Seq 98.43%)

4.2. Confirmation Test

Furthermore, the confirmation test verified the improvement of results and to predict the optimum performance at the selected levels (since all factors have a confident level more than 90%) of significant parameters such as A2, B3 and C2. Table 7 shows the comparison between the actual and experimental results. The predicted mean (M) values can be expressed as,

$$M = (A2-T) + (B3-T) + (C2-T) + T,$$

Where T= average of S/N ratio (Compressive strength) → 32.514

$$M= (33.29-32.514) + (33.10-32.514) + (32.86-32.514) + 32.514$$

(Compressive strength) M= 34.222 dB

Similarly, T = average of S/N ratio (Water absorption) → -12.091

$$M = ((-10.63)-(-12.091)) + ((-11.47)-(-12.091)) + ((-11.74)-(-12.091)) + (-12.091)$$

(Water absorption) M= -9.658 dB

The predicted mean of the response S/N ratio for the compressive strength lies in the range of 31.08 dB < Compressive strength > 34.22 dB at the confidence level of 95%.

The predicted mean of the response S/N ratio for the Water absorption lies in the range of -9.60 dB < Water absorption > -14.47dB at the confidence level of 95%.

Table 7. The comparison between predicted and actual results of compressive strength and water absorption

Output	Level	Response	Conformation	Experimental	Difference
Compressive strength	A2, B3, C2	Observed result (MPa)	51.45	49.66	1.79
		S/N Ratio in dB	34.222	33.92	0.302
Water absorption	A2, B3, C2	Observed result (%)	2.95	3.02	0.007
		S/N Ratio in dB	-9.658	-9.60	0.058

4.3. Regression analysis

The regression equation of compressive Strength = 42.536 - 5.966 A_1 + 3.724 A_2 + 2.241 A_3 - 1.453 B_1 - 1.650 B_2 + 3.103 B_3 + 0.141 C_1 + 1.788 C_2 - 1.929 C_3

The regression equation of water absorption = 4.0987 + 1.050 A_1- 0.675 A_2 - 0.375 A_3+ 0.061 B_1 + 0.198 B_2 - 0.259 B_3 - 0.061 C_1 - 0. C_2 + 0.217 C_3

5. Conclusions

In this present work the influence of nano SiO_2 in cement mortar were studied using the Taguchi technique. The following conclusions were drawn

- A2 B3 C2 combination represents 2 % nano SiO_2, W/C ratio 0.50 % and plasticizers 1 % obtain improved mechanical properties.
- From ANOVA table, it was observed that the effect of nano SiO_2 has a better improvement on compressive strength and water absorption.
- The highest maximum value for compression strength and minimum value for water absorption was found.
- The confirmation test evidenced that the optimum combination satisfied the actual requirement of process parameters to produce high performance cement mortar.

IOP Conf. Series: Materials Science and Engineering **992** (2020) 012013 doi:10.1088/1757-899X/992/1/012013

References

[1] Senff L , Hotza D, Repette W L, Ferreira V M and Labrincha J A 2010 *Construction and Building Materials*. **24** pp 1432–37.

[2] Shen W, Zhang C, Li Q, Zhang W, Cao L and Ye J 2015 Journal of Cleaner Production **87** pp 762- 765

[3] Sanchez F and Sobolev K 2010 *Construction and Building Materials* 24 pp 2060-71

[4] Said A M, Zeidan M S, Bassuoni M T and Tian Y 2012 *Construction and Building Materials* **36** pp 838-844

[5] Ltifi M, Guefrech A, Mounanga P and Khelidj A 2011 *Procedia Engineering* **10** pp 900-905

[6] Rahmani K, Shamsai A, Saghafian B and Peroti S 2012 *Middle-East Journal of Scientific Research* **12** 8 pp 1056 - 61

[7] Senff L, Labrincha J A, Ferreira V M, Hotza D and Wellington L Repette W L 2009 *Construction and Building Materials* **23** pp 2487-91

[8] Manzoor Z, Barbhuiya S, Shaikh F and Cheema D 2018 *Proc. of 24th Conf. on Mechanics of Structures and Materials: Advancements and Challenges* pp1871 - 74

[9] Liu X, Chen L, Liu A and Wang X 2012 *Int. Conf. on Future Energy Environment and Materials,Energy Procedia* 16 pp 991-996

[10] Mosaberpanah M A, Eren O and Tarassoly A R 2019 *Journal of Materials Research and Technology* **8** 1 pp 804- 811

[11] Bhanja S, Sengupta B 2002 *Cement and Concrete Research* **32**, 9 pp1391- 94

[12] ArunPrakash V R and Rajadurai A 2016 *Applied Surface Science* **384** (16) pp 99–106

[13] Murugan M A, Jayaseelan V, Jayabalakrishnan D. et al. 2019. *Silicon*

[14] Muthuramalingam T, and Mohan B 2013 *Indian journal of Engineering and Material science* **20** pp 471-475

[15] Jayakrishnan Unnikrishna Pillai, Ikshit Sanghrajka, Manikandakumar Shunmugavel T Muthuramalingam, Moshe Goldberg and Guy Little fair 2018 *Measurement* **124** pp 291-298

[16] Shoja M Z and Naddaf H E 2017 *Civil Engineering Journal* **3** 11

[17] Arunprakash V R and Viswanathan R 2019 *Composites: A* **118** pp317-326.

[18] ArunPrakash V R and Rajadurai A 2017 *Defence Technology* **13** pp 40–46.

[19] Zhang M H, Islam J and Peethamparan S 2012 *Cement and Concrete Composites*, **34** pp 650-662

[20] Natrayan L, Senthil Kumar M, Palanikumar K 2018 Mater. Res. Express **5** 066516

[21] Sharobim K G and Mohammed in H A 2013 *HBRC Journal* **9** pp 210-215

Enhancing the agro engineering system using game theory analytics

T P Anithaashri[1], G Ravichandran[2]
[1]Associate Professor, Department of Innovative Informatics, Institute of Computer Science and Engineering, Saveetha School of Engineering, Saveetha Institute of Medical and Technical Sciences, Chennai, Tamilnadu, India.
[2] Research Scholar, AMET University, Chennai, Tamilnadu, India

E-mail : shri3krra@gmail.com

Abstract: The advent of technologies and the impact of computer play an essential role in performing regular activities. The new era of Big Data analytics helps in analyzing the data through primary data and secondary data. By using modern tools to handle Distributed File systems enormous data can be handled and processed with high speed and accuracy in a stipulated time. The use of these emerging trends in all the fields gives better solution and strives the world in all aspects of digitization in a successful manner. To produce high yields in food crops, the agriculture adopted modern industrial systems. By using modern tools and techniques the agricultural production can be increased abundantly. The proposed system will help and attract the common people to invest their hard work and time in agricultural production, and make a huge contribution in the financial growth of the nation. The tractors and agro engineering tools for irrigation systems, are already used to improve the production of food crops. But the lack of advanced techniques such as smart system for monitoring the crop, to look for the strength of the soil through soil analyzer, checking for moisturisation in land and air, etc., are very much needed for proper yield. This system presents the novel framework and initial experimental results shows the feasibility of proposed system.

Keyword : Sensors, Big Data Analytics, Game Theory, Agro-Engineering

1. Introduction

Several Internet facilities and policies have been framed and put into practice by many governments all over the world in order to strengthen the agricultural fields. The basic need is to monitor all the parameters such as quality of soil earth we live-in, soil properties, weather predictions and this in turn alerts farmers, ranging from miniscule to large scale farmers to control these basic needs when required. In the year 2015 a policy framed by Ministry of Communication and Information Technology, India to transform digital landscape on Internet of Things and it was reported in the year 2019 to achieve superiority in the use of Information and Communication Technology (ICT) and to uplift our Indian factor to greater heights in the agricultural sectors. Australia, Ireland, France, China, Malaysia, USA, Thailand and Philippines are some of the other countries where IoT has been successfully implemented over the years from 2014 to till date.

The main objective of this proposed system is to inculcate the agriculture practices using innovative technologies and help in reducing global warming and make earnings for their healthy and happy life using smart Agro System. It helps in analyzing the soil, improvisation in irrigation system and monitoring

IOP Conf. Series: Materials Science and Engineering **992** (2020) 012014 doi:10.1088/1757-899X/992/1/012014

the crops using emerging sensor technologies and Big Data Analytics for processing and accessing the data quickly. Also by using Cloud Computing technology the communication can be shrunk with the hand held systems such as Mobile phones, i-pads, etc., which help the agriculturists to sell their yields in a profitable way.

2. Existing System

Agricultural field monitoring is a laborious and tiresome process[4]. It can be simplified and made profitable by reducing labor input and identify the solution within a few seconds. Recent advancement in ICT and data management[6],[7] made this possible with the help of computers and Internet facilities[1]. Several Internet facilities and policies have been framed and put into practice by Government of India in order to strengthen the agricultural sector[2]. The basic need is to monitor all the parameters such as quality of soil-earth we live-in, soil properties, weather predictions and this in turn notifies the farmers, ranging from miniscule[17] to large scale farmers to control these basic parameters when required.

Through the policy of India, to achieve superiority in the use of Information and Communication Technology (ICT)[10] and to uplift nation factor to greater heights in the agricultural sectors. Australia, Ireland, France, China, Malaysia, USA, Thailand and Philippines are some of the other countries where IoT [13]has been successfully implemented over the years from 2014 to till date. By not encouraging the irrigation transfer in many countries like Korea lead to lack water scarcity for the crop production. The aged farmers, lack of youth interest in rural livelihoods, unclear water-pricing structures support and lack of irrigation system and private investment, such social and demographic phenomenon has further eroded the self-governance capacity of rural communities

Most of the past surveys were done in water management, crop management, irrigation patterns, agricultural productivity, global population, effectiveness of the agricultural production, soil conditions such as pH, humidity, environmental temperature and soil moisture All existing systems, for example Object-oriented[15] expert system, Radial basis function networks, Rule based expert system, Artificial neural networks, Fuzzy Inference Systems, Image Processing, etc., concentrate on one or two parameters[14] at a time and their outputs are useful for cultivating[11], [12] only one crop [18] under similar weather conditions. Even if slight temperature variations in the parameters occur, the results [13] are not accurate to the extend desired. Also in most of the systems, the outputs are transmitted through internet facilities, processed and displayed in the monitor which is located indoor or within a nearby location.

To overcome these disadvantages there is an urgent need for an innovative application[8] that can be operated anywhere[12] around the country and monitor the agricultural lands [9][10]that are far from the reach of the farmers To obviate the above disadvantages, a Smart Agro based System is proposed. It has been initiated to develop a Smart Agro based system that can be installed in smart phones which will be a handy tool for small scale as well as large scale agriculturists.

3. Proposed System

The main objective of this proposed system is to inculcate the agriculture practices using innovative technologies and help in reducing global warming and make earnings for their healthy and happy life using smart Agro System[14], [16]. It helps in analyzing the soil, improvisation in irrigation system and monitoring the crops using emerging sensor technologies and Big Data Analytics for processing and accessing the data quickly. Also by using Cloud Computing technology the communication [2],[5] can be shrunk with the hand held systems such as Mobile phones, i-pads, etc., which help the agriculturists to sell their yields in a profitable way.

The development of this system involves three steps:

Step(1) : Smart Application for the Process of Soil Analyzes and Monitoring Irrigation Systems

Step(2): Monitoring the crops through Sensors

Step(3): Data Processing through Big Data Analytics

3.1. Smart Application for the Process of Soil Analyzes and Monitoring Irrigation Systems

The process of analyzing the soil for agriculture is a great challenge to the farmers. The proposed system will make a way to overcome this difficulty by processing the soil analyzes and monitoring the irrigation systems through smart applications. The monitoring of the agriculture systems helps the farmers to continuously aware of the quality of the soil (humidity, pH value, etc.,) and monitor the irrigation system.

3.2. Monitoring the crops through Sensors

The crops to be monitored using various kinds of sensors for the protection of crops from pests and worst climatic conditions, etc. By using multiple sensors, the signals indicate any mis-happenings in the agricultural field and thus the remedial actions could be initiated to protect the crops[18]. With the help of micro-programmed chips the received sensor signals are processed and appropriate actions initiated.

3.3. Data Processing through Big Data Analytics

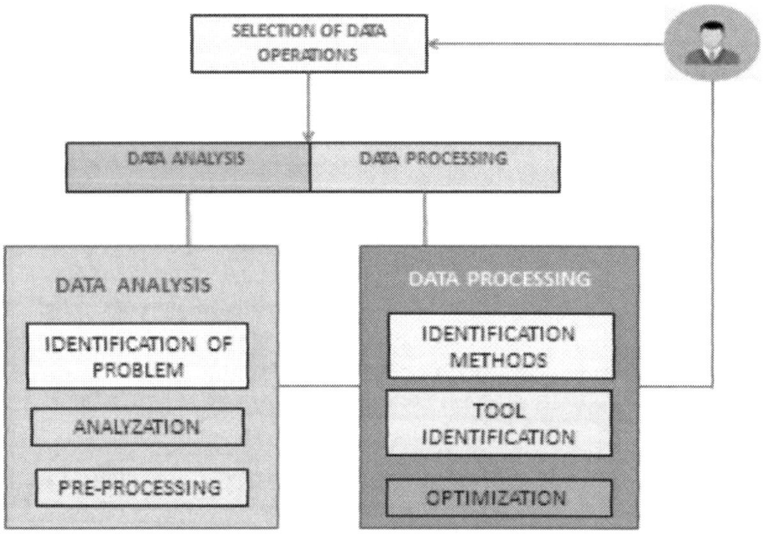

Figure 1. Smart System for Agro Engineering

IOP Conf. Series: Materials Science and Engineering **992** (2020) 012014 doi:10.1088/1757-899X/992/1/012014

The distributed system plays vital role in the data processing. By using the Avra and Sqoop in Hadoop Distributed File System so that the data are processed and accessed very quickly. The primary and secondary data are analyzed to predict the health of crops, quantum of production, make use of economical fluctuations profitably. The queries from various users relating to market rates are processed and responses can be provided spontaneously. In the Figure 1, the input will be given to the system and which identifies the problem and check the various factors[18], such as soil, weed control, climatic[14] condition, irrigation system, etc. After the identification of problem[17], it analyze and process the data to provide the accuracy in facts. Data will be analysed on the basis of principles for the equivalence of extensive and coalitional game strategies.

3.4. Elucidation of Game Strategies

An exchange factor that contains two agents (N = 2) and two goods have the preference relation, which can be conveniently represented as an payoff box[3]. Bundles of goods consumed by agent 1 are measured from one the origin in the bottom left, while bundles consumed by agent 2 are measured from the second origin in the top right. The width of the box formed by the two pairs of players is the total endowment of good stage in the factor. The more general notion of an factor with production.

Note that under our assumption that every agent's preference relation is increasing every agent is resource related to every other agent. Thus each point x in the payoff box corresponds to an allocation in which agent i receives the bundle x measured from set of players O* The point x* corresponds to a competitive allocation since the most preferred bundle of agent I in the set {Xi,pxi, pui} is x* when measured from origin O* using preference relation Ᵽ,

$$\text{pxi Ᵽ pui} > \text{Xi Ᵽ O*} \quad \text{----(1)}$$

The ratio of the competitive prices is the negative of the slope of the straight line through u and x*. An exchange factor wf is closely related to factor of a global market. Thus the model of an exchange factor as a coalitional game without transferable payoff has been framed. Precisely, it associate with the exchange factor (N, I, (wf), (Ᵽ-)) with the coalitional game (N, X, V, (Ᵽ{})) where

X = {(xi) i∈N: Xi Ᵽ R, for all i € N}; ------ (2)

V(S) = {(Xi) i∈N : Sxi ∈w* and xi = ui for all i in S} for each coalition game. ____ (3)

Each preference relation Ᵽ is defined by (xj)j€N, (yj)j€N if and only if x{} & y{}.

The third condition expresses the assumption that each agent cares only about his own consumption. Every competitive allocation in an exchange factor[8] is in the core. The idea behind the first solution concept that a coalition is unsatisfied with the current division of v(N) can credibly object by suggesting a stable division x of v(N) that is better for all the members of S and is backed up by a threat to implement (xj). The logic behind the requirement that an objection itself be stable is that otherwise the objection may unleash a process involving further objections by other coalitions, at the end of which some of members of the deviating coalition may be worse off. When making a choice at players last information set is not informed of the action of chance, though player was so informed and made with previous choice.

This strategic analysis helps to perform the predictive analysis on the class of the given data set. It also performs well in multi class prediction. The categorical input variables are much useful compared to numerical variable to analyze the data.

4. Results and Conclusion

Initially it has been tested with very few test beds, arrive a best enhanced results and shows the drastic improvement in predictive analysis for the good crop production. The table1.0 depicts the detection rate for various factors such as soil, water, disease, climate ad weed control. It shows enhanced results in time consumption to analyze the data approximately in 10 minutes less and makes the profitable at monetary level by increasing the maintenance of food crop production.

Table 1. Detection Rate of Analysis on various factors

Description	Detection Rate Normal	Detection Rate proposed system
Soil	90	84
Water	95	82
Disease	97	83.1
Climate	91	81
Weed control	92	83.7

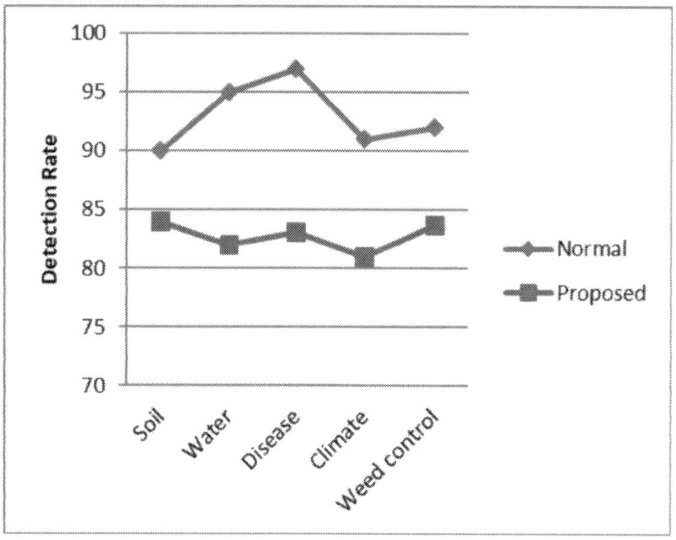

Figure 2. Detection Rate of Data Analysis

In future, to sell the food grains in the market, the mobile application will be launched so that the prices for the crops will be amicably fixed by the farmers. No distributors or brokers are entertained in the sale of the products. Thus the proposed system will be very useful for the agriculturists.

References

[1] Rupnik, R.; Kukar, M.; Vra˘car, P.; Košir, D.; Pevec, D.; Bosni´c, Z. AgroDSS 2019: A decision support system for agriculture and farming. Comput. Electron. Agric. 161, 260–271.

[2] Anithaashri TP, RavichandranG, Baskaran,R 2019 Software Defined Network Security enhancement using Game Theory, Elsevier COMNET, vol **157**, pp:112-121

[3] Shigen Shen and Longjun Huang and Haiping Zhou and Shui Yu and En Fan and Qiying Cao 2018 Multistage Signaling Game-Based Optimal Detection Strategies for Suppressing Malware Diffusion in Fog-Cloud-Based IoT Networks, IEEE *Internet of Things Journal*, vol. **5**, no. 2, pp. 1043 - 1054.

[4] Shuang Zhao and Yang Yang and Ziyu Shao and Xiumei Yang and Hua Qian and Cheng-Xiang Wang 2018 FEMOS: Fog-Enabled Multitier Operations Scheduling in Dynamic Wireless Networks, IEEE Internet of Things Journal, vol. **5**, no. 2, pp. 1169 - 1183

[5] Anithaashri TP, Ravichandran G , et.al. 2018 Secure Data Access Through Electronic Devices Using Artificial Intelligence, ICCES – IEEE Digital

[6] Kumar, A.; Sah, B.; Singh, A.R.; Deng, Y.; He, X.; Kumar, P.; Bansal, R.C. 2017 A review of multi criteria decision making (MCDM) towards sustainable renewable energy development. *Renew. Sustain. Energy Rev*, **69**, 596–609

[7] Pooja S, Uday D, Nagesh U,Talekar S, 2017 Application of MQTT protocol for real time weather monitoring and precision farming, in International Conference on Electrical, Electronics, Communication, Computer, and Optimization Techniques (ICEECCOT).

[8] Pandithurai O, Aishwarya O, Aparna B, Kavitha K,2017 Agro-tech: A digital model for monitoring soil and crops using internet of things (IOT) in Third International Conference on Science Technology Engineering & Management (ICONSTEM).

[9] Roselin A, Jawahar A 2017 Smart agro system using wireless sensor networks in International Conference on Intelligent Computing and Control Systems (ICICCS).

[10] Rekha P, Rangan V, Ramesh M, Nibi K,2017 High yield groundnut agronomy: An IoT based precision farming framework in IEEE Global Humanitarian Technology Conference (GHTC).

[11] Maia R and Netto I and Tran A 2017 Precision agriculture using remote monitoring systems in Brazil in IEEE Global Humanitarian Technology Conference (GHTC)

[12] Ahmad Z, Pasha M, Ahmad A, Muhammad A, Masud S et.al, 2017 Performance evaluation of IEEE 802.15.4-compliant smart water meters for automating large-scale waterways in 9th IEEE International Conference on Intelligent Data Acquisition and Advanced Computing Systems: Technology and Applications (IDAACS),.

[13] Mekala M, Viswanathan P, 2017 A novel technology for smart agriculture based on IoT with cloud computing in International Conference on I-SMAC (IoT in Social, Mobile, Analytics and Cloud) (I-SMAC)

[14] Qi D, Lu G et.al 2017 Design of Urban Greening Intelligent Monitoring System Based on Internet of Things Technology, in 9th International Conference on Intelligent Human-Machine Systems and Cybernetics (IHMSC).

[15] Dolui K, Datt SK, 2017 Comparison of edge computing implementations: Fog computing, cloudlet and mobile edge computing, Global Internet of Things Summit (GIoTS).

[16] Walter A Finger R,et.al 2017 Opinion: Smart farming is key to developing sustainable agriculture," *Proc Natl Acad Science*, vol. **114**, p. 6148–6150.

[17] Rose, D.C et.al. 2016 Decision support tools for agriculture: Towards effective design and delivery. *Agric. Syst.* **149,** 165–174
[18] Zhang Q 2015 Precision Agriculture Technology for Crop Farming, 1st ed.; CRC Press and Taylor & Francis Group: Boca Raton, FL, USA ISBN 978-1-4822-5107-4.

IOP Conf. Series: Materials Science and Engineering **992** (2020) 012015 doi:10.1088/1757-899X/992/1/012015

Novel approach to enhance network security using key performance indicator in business analytics

M.Saranya[1], T.P.Anithaashri [2]

[1]Assistant Professor, Department of Innovative Informatics, Institute of Computer Science and Engineering, Saveetha School of Engineering, Saveetha Institute of Medical and Technical Sciences, Thandalam, Chennai, Tamilnadu, India-602105.
[2]Associate Professor, Department of Innovative Informatics, Institute of Computer Science and Engineering, Saveetha School of Engineering, Saveetha Institute of Medical and Technical Sciences, Thandalam, Chennai, Tamilnadu, India-602105.

E-mail : Saranya.cse01@gmail.com

Abstract: In Recent days, the challenges in Business applications are more in analyzing the complexities of day to day activities. Key execution pointer (KPIs) are business estimations used by corporate bosses and various directors to follow and examine factors thought about noteworthy to the achievement of a Business. Reasonable KPIs base on the business cycles and limits that senior organization sees as commonly huge for assessing progress toward meeting imperative destinations and execution targets. KPIs differ from relationship to affiliation subject to business needs. One of the key display pointers for an open association will likely be its stock expense, while a KPI for a covertly held startup may be the amount of new customers incorporated each quarter. Surely, even direct adversaries in an industry are likely going to screen different plans of KPIs specially designed to their individual business procedures and the leader's strategies for thinking.

Keywords: Security Metrics, Balanced Score Card, Quality of Service, KPI.

1. Introduction

1.1 Network Security

Computer Network Security [1] comprises of measures taken by a business, association or other establishment to screen and forestall unapproved access or abuse of a Computer organization and its assets. The executive's approaches have various necessities, contingent upon the size and physical setting of the organization. A home or little office may just require essential Computer network security, while huge organizations may require high-upkeep and progressed programming and equipment to keep assaults from hacking and spamming. Admittance to data and projects on the organization are constrained by a Network Administrator, who appoints clients an ID and secret word or other verifying data to guarantee that they are approved to direct exchanges and interchanges over the organization.

1.2 Cyber Security

Network safety is important to productively work the business but at the same time it's basic for securing [2]the clients data.In IT best practices goes past utilizing the most recent innovation at the company. So it's useful to utilize a layered security model.

1.3 Network Access Control

Network Access Control (NAC) is a way to deal with PC security that endeavors to bind together endpoint security innovation, (for example, antivirus, have interruption counteraction, and weakness evaluation), client or framework confirmation and organization security authorization. Network Access Control (NAC) is an organizing arrangement that utilizes a lot of conventions to characterize and execute a strategy. Network Access Control (NAC)[10] is primarily utilized for endpoint wellbeing checks, yet it is regularly attached to Role-based Access. Admittance to the organization will be offered by the profile of the individual and the consequences of a stance/wellbeing check. For example, in an Undertaking the HR division could get to just HR office records if both the job and the endpoint meets against infection essentials.

2. Related Works

2.1 Intrusion Detection System

Protect the business [2] from destructive and dubious organization movement through interruption discovery frameworks (IDS) and interruption avoidance frameworks (IPS)[3] as "a security apparatus that screens network traffic looking for dubious action and known dangers." An IDS screens organizations and gadgets to reveal noxious or hurtful action and send alarms when it discovers likely dangers. An interruption identification framework is intended to distinguish potential dangers and weaknesses against a Computer. This is a fundamental component of ensuring the business and store so much significant data on the Computer.

2.2 Encryption

Encryption [8] is a troublesome idea to get a handle on, however it's a vital piece of ensuring the business information. The reason for document and circle encryption is to ensure information put away on a PC or organization stockpiling framework. All associations, including little and medium size organizations (SMBs), that gather by and by recognizable data like names, birthdates, Social Security[9] numbers and monetary data must make sure about that data.

2.3 Network Monitoring

Monitoring [4] is the most ideal approach to keep a heartbeat on the exhibition of the business' organization. The scope of measurements, including transmission capacity use, misfortune, jitter, inertness

3. Proposed System

3.1 Key Performance Indicator

A Key Performance Indicator [12] (KPI) is a kind of measure that is utilized to assess the exhibition of an association against its key destinations. KPIs can be utilized to follow the operational presentation of offices, tasks or people against targets or objectives

3.1.1. Lagging KPI

Lagging KPIs [13] are utilized to decide the consequence of past execution, for example,

Creation, volume or an outcome.

3.1.2. Leading KPI

Leading KPIs are utilized to foresee or impact future execution. They are more hard to set up as they depend more on outside activities to affect results, for example, changes in cycle or interests in framework.

3.1.3. Operational KPI

Operational KPIs plainly articulate detailed and ideal data that is utilized to settle on everyday choices, or take remedial activities on execution or a cycle.

3.1.4. Strategic KPI

Strategic KPIs [14] are centered around long haul destinations got from an associations objectives.

3.2 Architecture Diagram

The Balanced Scorecard is a notable execution the board device. It recommends following a circumstances and logical results rationale to construct a tree of business destinations moving from a money related viewpoint down to clients, cycle and instruction and development, and afterward improving the whole plan with KPIs [16] and activity plans.

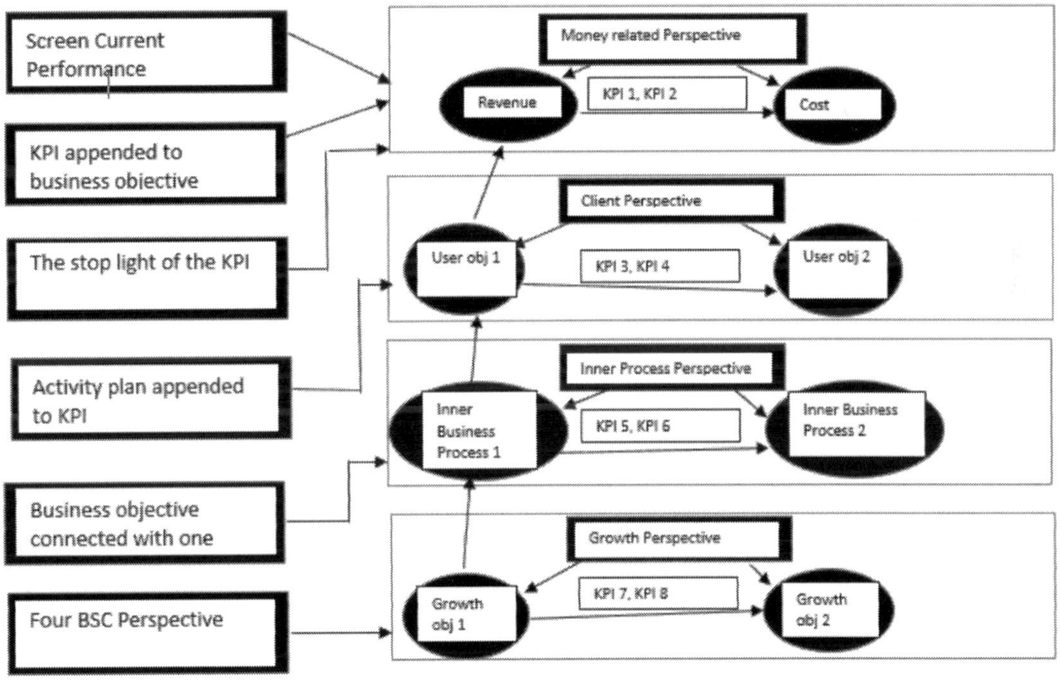

Figure 1.Balanced Score card with KPI

Financial measures pass on the financial consequences for the moves previously made by the association, and spotlight on the benefit related measures. On this client perspective catches the capacity of the association to give quality merchandise and enterprises, the viability of their conveyance, and generally speaking client care and fulfillment. This will result from value, quality, accessibility, determination, usefulness, administration, associations and brand esteem suggestions. Internal measures the interior business results that lead to monetary achievement and fulfilled customers. These key business measures are observed to guarantee that results will continuously be palatable. The interior cycle's point of view investigates the proficiency of inside cycles and techniques. (Figure 1)

3.3 Network Traffic

Network traffic [10] or information traffic is the measure of information moving over an organization at a given purpose of time. Organization information in Computer networks is generally epitomized in network bundles, which give the heap in the organization.The workers could be depleting network execution without acknowledging it. In the event that the workers utilizing transfer speed hefty applications, web based video or music, or downloading a few enormous records through the organization, there'll be significantly less traffic [7] accessible for basic

Business tasks.[5] It's significant that your undertaking speaks with its clients about appropriate utilization of the organization and which applications/administrations shouldn't be utilized on a business organization.

3.4. Balanced Score Card

A Balanced scorecard [11] is a methodology execution the executive's tool – a semi-standard organized report that can be utilized by chiefs to monitor the execution of exercises by the staff inside their control and to screen the outcomes emerging from these activities.

To follow KPIs, most organizations utilize a Balanced Scorecard. Adjusted scorecards have for some time been utilized in vital business the executives to follow key execution pointers, and are intended to give a structure to oversee assets. There are four fundamental perspectives or viewpoints to take with

The KPI adjusted scorecard: [12] Budgetary viewpoint – following monetary execution. Client point of view – following consumer loyalty, perspectives, and piece of the overall industry objectives. Inner cycle viewpoint – covers inside operational objectives expected to meet client destinations.

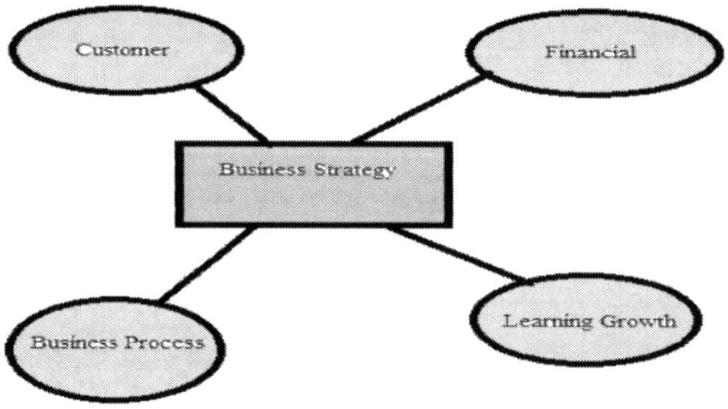

Figure 2.Key Performance indicator and Balanced Score card

Explaining and making an interpretation of vision into procedure – distinguishing key destinations and catching them in a vital guide. This requires the business chiefs to arrive at agreement on what the vital destinations really are, and that all goals are connected. Imparting key [15] goals and gauges and connecting them to tasks – this incorporates building up a two-path interchanges with partners to urge exchange to refine measures and advance criticism. Arranging and setting focuses to adjust key activities – including recognizing focuses of every one of the targets estimated by the KPIs. In a perfect world, singular targets are set for key destinations and tasks,[16] which are thus connected to the bigger key

IOP Conf. Series: Materials Science and Engineering **992** (2020) 012015 doi:10.1088/1757-899X/992/1/012015

goals. Improving vital input and taking in – this incorporates gaining from execution data and utilizing the discoveries from the reasonable scorecard to refine dynamic.(Figure 2)

3.5. Security Metrics

A Business Metric [5] is a quantifiable measure that is utilized to follow and evaluate the status of a particular business measure. Business measurements [6] ought to be utilized to address key crowds encompassing a business, for example, speculators, clients, and various sorts of workers.

4. Results and Conclusion

The learning and development point of view looks at the capacity of workers (aptitudes, abilities, information and preparing), the nature of data (frameworks, information bases and networks) and the impacts of authoritative arrangement (culture, authority, arrangement and cooperation), in supporting the achievement of authoritative objectives".The collection of information examination is really to do information correlation investigation. Without information examination, it is regularly hard to utilize single marker measurements to play the estimation of information shown in Table 1.

Table 1. Balanced scored card with KPI Analysis

SI.NO	Objectives	Target	Current	To Target
1	Money point of view	4500	5700	25%
2	Client Perspective	3200	2900	-10%
3	Inner target Perspective	2500	1000	-40%
4	Learning and development point of view	100	70	-30%

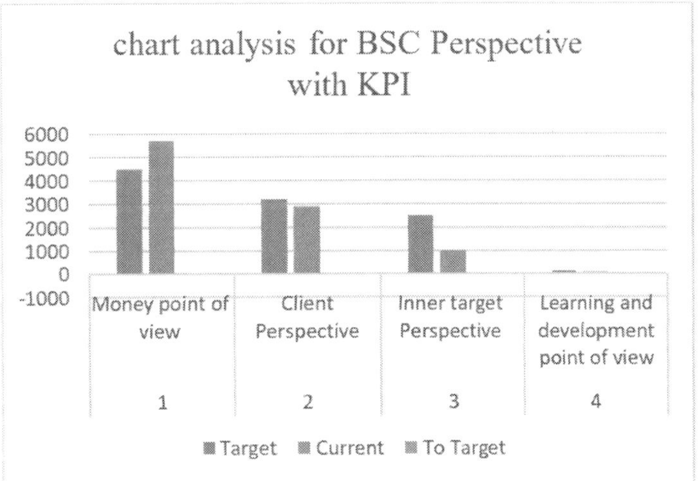

Figure 3. Analysis for BSC Perspective with KPI

The paper exhibits various methodologies for investigating key execution markers in various ways including manual, choice, or forecast draws near. As there are various thoughts, nonetheless, observed that the expected approach is as yet an indispensable field for research as a large portion of the examination depend on a decided point. The future bearing after this broad exploration audit is to propose another methodology for foreseeing KPI's with no decided highlight begin and give a total expectation to the KPI's including the appropriate KPI's for the issue and their conceivable worth range. (figure 3).

References

[1] William Nugent 1985 Methods for measuring the level of computer security. *NIST Special Publication* **500-133**

[2] Reiter M.K and Stubblebine.S.G 1999 Authentication metric analysis and design.*ACM Transactions on Information and System Security*, 2(2):**138–158,5**

[3] Nielsen F, 2000 Approaches to security metrics," in *CSS-PAB Workshop* on" Approaches to Measuring Security,

[4] Andersen, H., Cobbold, I., & Lawrie, G. (2001). Balanced scorecard implementation *SMESME Copenhagen, Denmark.*

[5] Swanson M, Bartol N, Sabato J, Hash J, and Graffo L. 2003 Security metrics guide for information technology systems. *NIST Special Publication* **800-55**,

[6] Chapin, D. A. & Akridge, S. (2005).Information Security Measurement, *Information Systems Audit and Control Association.*

[7] J. Pamula, S. Jajodia, P. Ammann, and V. Swarup.2006 A weakest-adversary security metric for network configuration security analysis. *In Proceedings of the 2nd ACM workshop on Quality of protection*, pages **31–38**

[8] Kadarsah Sordid 2007 Key performance indicators measurement model based on analytic hierarchy process *International symposium on the analytic hierarchy process (ISAHP)*, Mar, Chile, vol. **3**

[9] John Gilligan 2009. Information technology Security techniques *International Organization for Standardization.*

[10] Rainer Bohme, 2010"Security Metrics and Security Investment Models",*International Computer Science Institute, Berkeley, California, USA*

[11] Deepti Juneja, Kavita Arora, Sonia Duggal, 2011"Developing Security Metrics for Information Security Measurement System", *International Journal of Enterprise Computing and Business Systems*, Vol. 1 Issue **2**

[12] Abdel-Kader M., Moufty S., & Laitinen, E. K. 2011. Balanced Scorecard Development: *Literature and Directions for Furture Research.*

[13] Ingo Keck R, Robert Ross J 2014 Exploring customer specific KPI selection strategies for an adaptive time critical user interface *Proceedings of the 19th international conference On intelligent user interfaces.*

[14] Zhou Ning, Wang Jian-hai, Wu Jia-xin 2011 Design study on human-computer interface in Kpi system of enterprises *Applied Informatics and Communication Springer Berlin* pp. **189-195**

[15] El-mongy A.M.A, Hamouda A.E, Nounou N. Wahdan A.A 2013 Design of prediction system for key performance indicators in balanced scorecard *Int J Comput Appl*, **72 (8)**

[16] Mohamed Khalifa, Parwaiz Khalid 2015 Developing strategic health care key performance indicators: *a case study on a tertiary care hospital Comput Sci,* 63 pp. **459-466**

IOP Conf. Series: Materials Science and Engineering **992** (2020) 012016 doi:10.1088/1757-899X/992/1/012016

CNN based framework for identifying the Indian currency denomination for physically challenged people

P.Selvi Rajendran[1] , Dr. T.P. Anithaashri[2]

[1]Hindustan Institute of Technology and Science, Chennai

[2]Institute of Computer Science and Engineering, Saveetha School of Engineering, Saveetha Institute of Medical and Technical Sciences, Chennai - 602 105, India.

Email: selvir@hindustanuniv.ac.in

Abstract: One of the premier issues confronting the visual hindered individual is money, acknowledgment especially for cash. Be that as it may, the outwardly weakened individual may not think about the estimation of cash, and they endure part in cash trade related issues in their normal life. To address this issue, we have built up a framework for acknowledgment of money, notes, which might be the helpful device for an outwardly debilitated individual. Investigation and trials were done on the money informational collection, which encouraged CNN dependent on the key highlights, for example, watermarks, pictures printed on cash, esteemed composed as words and numbers and the total cash. This paper deals with the utilization of Convolutional Neural Networks (CNNs) for solving this society issues and investigations about the exhibition and evaluation of different CNN models. Here, Alexnet, Googlenet, and Vgg16 models have been considered for assessment. All the models were adjusted as far as preparing and testing the individuals of data sets. Among these three models, Alexnet accomplished better in preparing fulfillment, Vgg16 model indicated the better execution and accomplished 100%, Google net arrives at 88% as far as productivity.

1. Introduction

According to the Indian authorities, the big proportion of visually impaired people became higher [1]. About one hundred and sixty-five people were visually impaired in line with lakh individuals. Blind people were eighty-two percent of them, and low vision was 18 percent [1]. The latest smartphone development makes it possible to identify currency with an attractive one. In this research work we apply the convolutional neural network models such as Alexnet,Googlenet and VGG16 to detect the value of the currency notes.

In the present situation, the reputation for currency denomination is becoming a vibrant topic for scientists in multiple apps such as blind person currency recognition and fake currency detection for normal people and apps linked to transactions. An important aspect of our normal life is money transactions. Each nation has a distinct currency layout and value. The distinctive skills, colorations, denominations, and the global value for their currency have been retained by each individual county. The denomination can be acknowledged for foreign cash without trouble, but from the real currency, it will become difficult to interpret a fake currency. Particularly suffer in cash exchanges for the visually impaired people. They may not now be ready to properly comprehend one-of - a-kind denominations and are often betrayed by distinct individuals.

We speak to an application that distinguishes cash by means of an application and dispatches the result by means of sound frameworks. One of the essential issues confronting outwardly weakened individuals

IOP Conf. Series: Materials Science and Engineering **992** (2020) 012016 doi:10.1088/1757-899X/992/1/012016

are the failure to see paper monetary forms because of the guess of the paper surface and span between excellent monetary and standards. Consequently, this present framework's capacity is to build up a disposition to taking care of this issue so as to cause dazzle people to feel safe and take care of the financial issues

Regarding picture inclusion, there are two critical fields of cash acknowledgment (i) scanner-based methodology and (ii) camera-based methodology. The scanner-based system takes a picture of the report all in all (like a scanner). These frameworks are proper for money counter offices. At the same time with camera-based methodology catching the paper utilizing a versatile camera that can be part of the report. The scanner-based technique is tended to by the most extreme related writing work[2-5]. It is affirmed to permit clients to catch any portion of the currency note by their advanced smartphone and permit the framework to inform the expense of the cash to visually disabled persons for the utilization of unmistakable weakness of blind individuals.

Right now, based Indian paper money is educated to analyze the utilization of profound learning models that have the picture preparing instruments that could make taking care of time extremely quick with the right exactness. The proposed structures are fit for taking care of in part and under unmistakable lighting conditions caught records.

Profound learning procedures are broadly utilized in applications proper for picture acknowledgment nowadays. Convolutional Neural Network(CNN) is one among the notable systems that were esteemed one of promising strategies in issues of picture recognition[11,12,13]. The accuracy of the calculation for cash acknowledgment can be drastically used for training with a major number of data. Subsequently, acknowledgment can be performed well on a progressing premise and accomplishes high accuracy relying upon the measure of picture information allotted for preparing. For cash paper, incalculable methodologies have been performed.

Utilizing distinctive convolution neural system models, camera-based Indian paper cash is recognized on this exploration work, making preparation time extremely short with alluring exactness contrasted with the previous techniques referenced in the writing study. The recommended procedure can treat the incompletely and under unmistakable light conditions caught fiscal reports.

The rest of the paper is organized in the following way; Section 2 presents work of writing connected to consider research on cash recognition and versatile applications. The subtleties of the proposed framework and techniques are discussed in Section 3. The details of the data set and the collection of currency and coin data set describes in section 4. In section 5 explains popular CNN models available for image processing. Results and analysis are discussed in section 7.

2. Related Works

This section discusses works associated with literature. In recent years, various bank currency identification strategies or algorithms have been used to recognize and classify banknotes from various nations.

Computer vision: The writers acknowledge currencies based on computer vision in their approach [6]. They intended a scheme for recognizing four distinct currencies using extraction methods of characteristics. The main characteristic considered were currency texture, color, and forms. For classification, they used Artificial Neural Network. The median level of accuracy was 93.84%.

SIFT Algorithm: Iyad and their team created a visually impaired portable currency identification scheme and used Jordanian currencies as a test dataset [2]. Their strategy is based on the algorithm of the invariant transform function (SIFT) scale [10]. It attained the accuracy level of 71% to recognize the currency and 25% for the coin type of currency

Radial Basis Function Network: The author has created a portable paper currency identification scheme in [15,19] and has been tested on Saudi Arabia documents. The currencies were recognized here based on picture characteristics and correlations between two pictures. For classification, it utilizes the Radial Basis Function Network. The scheme is 95.37 percent accurate for Normal Non-Tilted Images, 91.65 percent for Noisy Non-Tilted Images, and 87.5 percent for Tilted Images.

Matching Template: Sungwook and their team have created an algorithm [4,18] which can recognize currencies from different countries. This technique identifies currencies based on the distinct currency note size and the multi templates correlation matching value. They achieved 100% precision in classifying and identifying the currencies of five distinct countries.

Image processing: Neoura has proposed the strategy which used image processing characteristics to define the currency [7,8,17]. Image segmentation, equalization, region of interest (ROI) extraction and then matching the image based on template, correlations between the images taken and the database dataset were the primary operations applied to the picture on their research work. This strategy has accomplished the precision of 89 percent in recognizing the currency. The information set for the Egyptian currency was used to test the scheme.

LBP Algorithm: Junfang's approach identified the currency based on the characters extracted from the image [14,16] and used the Local Binary Pattern approach to process the image. It achieved the precision of 92 percent.

In this paper, convolution neural networks are used for the classification of bank currency notes, labeling the boxes into which the input image is divided into in the network, in this case 11 X11 layers are used. Each cell has the probability of being one of the 5 or 7 classes of Indian banknotes and Coins, respectively, and the network also performs a segmentation of the object or banknote, enclosing it with a bounding box in the image.

3. Proposed Approach

This work presents a system to detect and recognize banknotes from different denominations using convolutional neural networks with deep training. It involves the following processing steps.

1. The proposed convolution neural network is trained using large collection of images taken in the different lighting condition such as natural and artificial lighting effects, with variations in the lighting and different rotation, Here, (x, y) coordinates represents the predicted bounding box position for the given input image and $(\hat{x}, \hat{y})$ represents the actual or original positions from the training data.

2. To deal with the error in deviation among bounding box and actual marked place, instead of calculating the width and height directly the square root of the predicted bounding box width and height is calculated.

$$\sum_{i=0}^{s^2} \sum_{j=0}^{B} \prod_{ij}^{obj} \left(C_i - C_i\right)^2 + \lambda_{noobj} \sum_{i=0}^{s^2} \sum_{j=0}^{B} \prod_{ij}^{obj} \left(C_i - C_i\right)^2 \quad (1)$$

Loss function -I

The parameter λ represent here to utilize the initial segment contrastingly and to assign the weight portions of the misfortune capacities. This is important to expand the soundness of the model. The most noteworthy punishment is for arranging forecasts (λ coord = 5) and the least for certainty expectations when no item is available (λ noobj = 0.5).

3. Loss related with the certainty score for each bounding box indicator is computed. C is the certainty score and ^ C is the Crossing point over Union of the anticipated bounding box with the ground truth. 1 obj is rise to one when there's a question within the cell, and something else. 1noobj is the opposite.
4. Another portion of the loss function is the object classification loss. The term 1noobj is utilized to penalize a classification error when no object is show within the cell.

$$\sum_{i=0}^{S^2} \mathbb{1}_i^{obj} \sum_{c\in classes} (p_i(c) - \hat{p}_i(c))^2 \quad (2)$$

Loss function -II

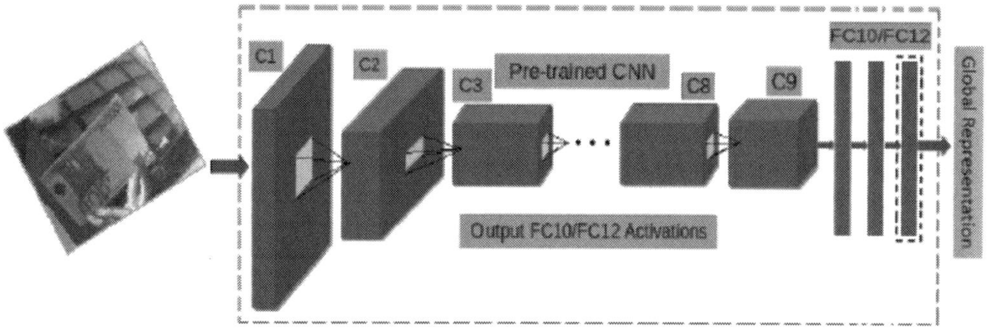

Figure.1 Pre-trained Model of the Convolutional Neural Network.

The following step was carried out during CNN Training.

1. First, the CNN model should be pre-trained with the primary 20 convolutional layers.The ImageNet 1000-class dataset has been used for training with a input measure of 224x224
2. In the next step, input resolution of the image has been increased to 448x448
3. The complete network has been trained for the rate of parameter setting are almost 135 epochs employing a batch measure of 64, momentum of 0.9 and decay of 0.0005
4. Learning rate plan: for the primary epochs, the learning rate was gradually raised from 0.001 to 0.01. Train for almost 75 epochs and after that begin diminishing it.
5. Use information expansion with arbitrary scaling and interpretations, and arbitrarily altering presentation and saturation.

This framework also follows the same method of using the rectified linear units (ReLUs) through all of the CNN structures, and the completely-connected layers. This activation feature is popularly utilized in CNN because it works as a rectifier that lets in the simplest advantageous values to skip and enables to enhance the training pace as well as keep away from the gradient-vanishing, as compared to sigmoid function

Figure. 1 suggests the model of the network architecture used for the recognition of Indian banknotes. The network used to categorize banknote denominations consists of nine convolutional layers accompanied by 2X2 Maxpool layers, 3 fully-connected layers and an exit or detection layer. This structure is based totally on and trained in the darknet framework. The memory used for every photograph is about 34.6 MB and the Parameters for each photograph for ahead propagation by myself is about 46.5 million.

The output of this version is a 7 X 7 X 15 tensor for the Indian banknote community and a 7 X7 X 17 tensor for the Indian coin network. This tensor can be considered a 7 X 7 grid representing the input picture, where each tensor cell may have the two field definition. Every container carries the parameters of x, y, width, top, and the confidentiality of items belonging to the lessons.

The network was trained with the hardware components of NVIDIA Geforce GT 740 SC graphics card (2GB VRAM), set up in a laptop with a i7 three.50 GH central processor with 8GB in RAM for approximately Five days.

4.Data Set Acquisition

The different categories of Indian currencies differs in value estimation and color usage, separated from the quality of printing ,material used for printing and other which makes for simple visual distinguishing proof. In any case, for the visually disabled person, the content and color will not give the assistance at all and measure can lead to disarray since of the comparable measurements of the different coins. There are no benchmark datasets are accessible, due to this, the modern dataset has been made which comprises of 150 pictures of the currency note and coins. The 20% of the dataset is utilized as the test set for fine-tuned CNNs models, i.e. Alexnet, Googlenet, and Vgg16. All the dataset contains pictures of the front and the back sides of the Indian cash notes: 5, 10, 20,50,100,200,500 and 1000 rupees, in expansion to 1, 2, 5, 10, 25 paise, 50 paise and 1 rupee coins, 10 pictures have collected of each denomination. Figure 2 and 3 shows the picture of the dataset.

Figure.2 Sample data of Indian Rupees

The above Figure 2. Shows different pictures of Indian currency dataset which includes the backside and front side of the various Indian currency denomination images. All the pictures have been taken using high configured cameras, with high resolutions like VGA, 1.3 megapixels (MP), 2 MP and 5 MP, are the default settings of the cameras. For each cycle, there are 4 different half folds and 2 full-length setups was made. For each group we consider at slightest 12 different currencies, over 6 distinctive indoor situations and 7 different open air situations, whereas collecting the dataset. This presents numerous varieties in brightening, foundation and posture within the dataset. The dataset contains pictures of both modern and old currency, as well as currency with writes on them.

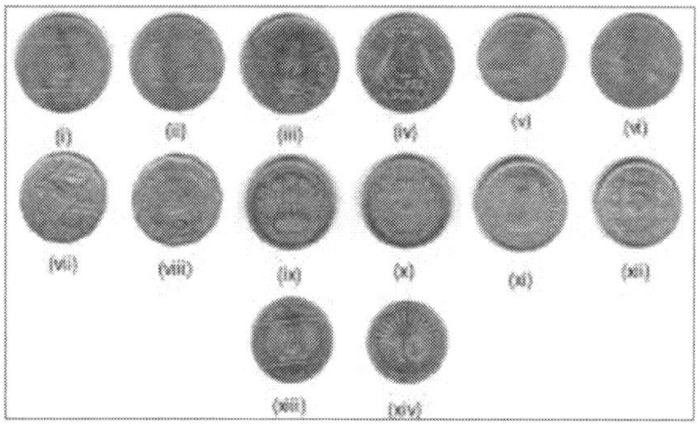

Figure.3 Sample Indian coins

5. Popular CNN Models

Generally, A Convolutional Neural Network prepare (CNN) fashions is the mixture of multi-layer neural structures, and it's far committed to recognizing the image highlights of pixel making utilize the shape of their pixels. The slicing edge and normally utilized CNN fashions are AlexNet, VGG16/19, GoogLeNet, and ResNet. In this section, the most pertinent models to recognize the currency, AlexNet, GoogLeNet, and VGG 16 have been taken into consideration for getting ready and testing the facts.

5.1 Alexnet

AlexNet is a form of convolutional neural network and it is already well trained in the millions of various categories images from the ImageNet database [1].This network structure have become one of the first deep neural networks to drive ImageNet magnificence accuracy through a big stride in evaluation. Its architecture framework composed of five convolutional layers accompanied through three completely connected layers, as depicted in the figure 1.

This type of network will have 8 deep layers and it may classify the input images into 1000 of categories, which include keyboard, mouse, pencil, and plenty of animals. As an end result, the Alexnet has observed out wealthy function representations for an extensive kind of input images. The network has a picture input duration of 227 by 227.

This Alexnet shape has changed into one of the initial level deep neural networks to classify the image net class accuracy by way of the use of a big stride in evaluation to traditional methodologies. It's far protected five convolutional layers followed by using 3 dedicated layers, as demonstrated in the figure 4. The depth of the networks is eight layers deep and may classify photographs into one of the thousand classes, in conjunction with keyboard, mouse, pencil, and lots of animals. As a result, the network version has educated and determined the rich feature illustration of a large variety of images. The length of the architecture is 227 by 227.

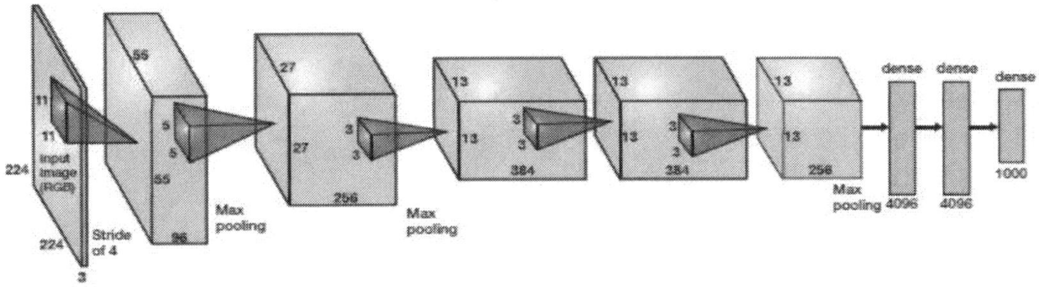

Figure.4 Architecture of Alexnet

5.2 Google Net

This network also has the similar shape of CNN version and it used a CNN inspired by way of LeNet however executed a completely unique detail that is dubbed an inception module. It used batch normalization, image distortions and RMSprop. This module is based on numerous very small convolutions as a way to substantially lessen the quantity of parameters. Their form consisted of a 22 layer deep CNN but decreased the large form of parameters from 60 million (alexnet) to four million

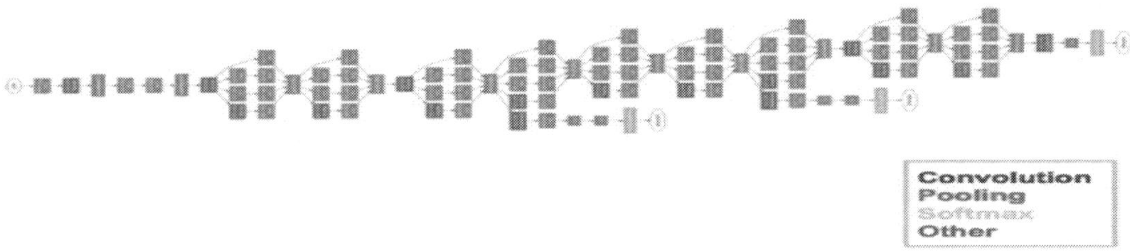

Figure. 5 Architecture of Googlenet

5.3 VGG Net

This model is made up of sixteen convolutional layers and may be very appealing due to its keeping the very uniform structure of some of the layers. Just like AlexNet, this network additionally has most effective 3x3 convolutions, however masses of filters. The weight configuration and parameter values of the VGGNet is publicly available for research purpose and has been utilized in lots of various applications and challenges as a baseline characteristic extractor. However, VGGNet consists of 138 million parameters, which may be hard to address.

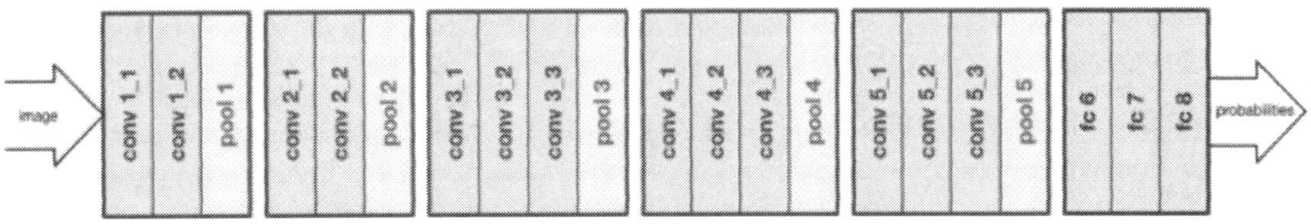

Figure.6 Architecture of VGGNet

5.4 ResNet-50

ResNet-50 is also the kind of convolutional neural network that is trained on more than 1,000,000 pics from the ImageNet database [1]. This type about 50 layers deep and can classify images into one thousand item classes, which incorporates keyboard, mouse, pencil, and masses of animals. As a result, the network has found out wealthy characteristic representations for a big variety of pictures. The network has a photograph input duration of 224 by 224.

5.5 Fine-tuning

The closing three layers of the CNNs which are already pre trained are retained as a fixed characteristic extractor for our dataset. The first step in switch mastering is mainly for replacing the last 3 layers in the pre-educated CNN with a hard and fast layer that may classify our targeted 11 instructions.

6. Results

This segment gives points of interest around preparing and testing the three fine-tuned models. To begin with, Figure 7 appears to be performance-based precision for preparing Vgg16, Alexnet, and Googlenet separately. It clearly appeared that Vgg16 illustrates the most extreme execution compared to Alexnet and Googlenet. To differentiate, Vgg16 required a longer time to total the preparation. In terms of learning speed, Googlenet shows up to be speedier than Alexnet and Vgg16 as outlined in Figure 7. This figure gives more subtle elements approximately the fine-tuned CNNs misfortune work.

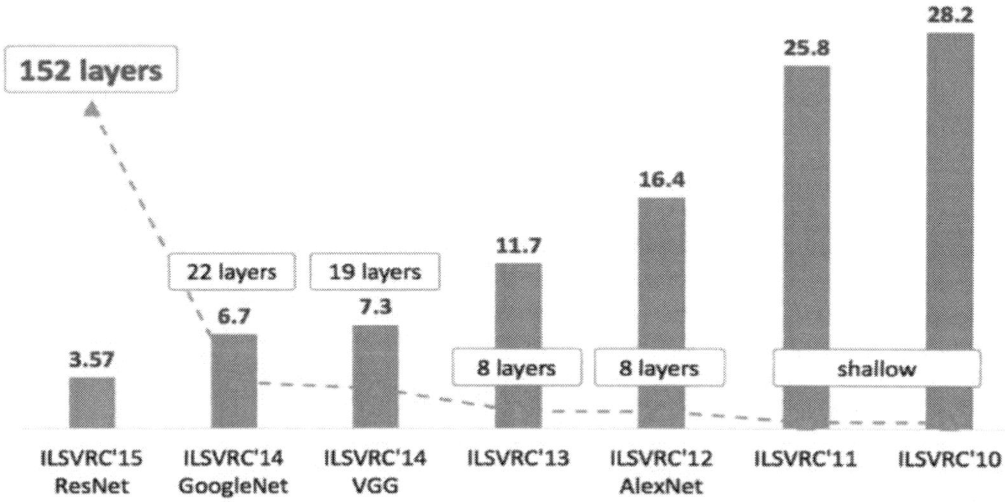

Figure.7 Classification performance of AlexNet and GoogLeNet trained

6.1 Experimentation

In this section, we first introduce the experiment procedure after which, talk the outcomes to prove the robustness and sensitivity of our approach. In this connection, the android utility has been designed for visually affected customers, there may be no symbols or buttons and no guide preparations needed. The application starts perusing the database of the spare cash banknotes at that point and begins taking pictures of the fronted scene. Once both flat edges are recognized a vibration effect takes place to tell the customer about the cash detection. After a while, a sound articulates the proper esteem in english is listen. Figure 8 appears to be the application running on Samsung GT-N800 tablet testing Rs.2000/- cash paper.

6.2 System Evaluation

This developed framework has been assessed using Matlab software tools based on accuracy measurement. The formula which is used to evaluate the system are

$$Accuracy = \text{True}/True+False \times 100 \quad (3)$$

In the above equation, true represents both genuine positive and genuine negative comes about whereas untrue shows the inverse. Table 1 appears to be the accuracy of our framework calculated for the 120 test data pictures (20 tests for each category). The main observation from table 1. are the highest accuracy obtained is 96% and it achieves to recognize the currency value 20 and 1000. The lowest accuracy comes about for the 50 and 500 rupees. It's gathered that this value was brought about due to the truth that within the 50 currency paper, the beat right corner has exceptionally few points of interest which may cause the blame proportion. The normal running time for Matlab framework is 10 seconds whereas in android framework is 12 seconds.

Table 1.Currency Detection Accuracy

₹	**5**	**10**	**20**	**50**	**100**	**200**	**500**	**1000**
Quantity	50	50	50	50	50	50	50	50
True	46	47	48	44	46	47	44	48
False	4	3	2	6	4	3	6	2
Accuracy(%)	92	94	96	88	92	94	88	96

7. Conclusion

In this paper, Indian cash recognition framework has been proposed. The proposed framework begins with capturing still pictures. Basic picture handling strategies like thresholding, noise evacuation, histogram equalization and division are utilized to extricate the ROI and encourage the layout coordinating method. Correlation based format coordinating is utilized after that to discover the ROI within the dataset pictures. This system has been developed in Matlab and OpenCV library beneath Android stage as well. The Matlab framework considers offline captured pictures whereas the Android one was planned to coordinate visual disabled clients. Live video capturing is considered for the Android application. This paper furnished a proof of the utilization of the idea transfer getting to know for Banknote recognition. Indian currency papers have been used for training and testing to tune the network models, namely Alexnet, Googlenet model, Vgg16 confirmed an terrific performance compared to Alexnet and Googlenet. While Vgg16 achieved the 100% accuracy whilst it is ninety five percent achieved by Alexnet and Googlenet attains 88% of accuracy.

In future, after making the final model which already trained with datasets, an app will be made and embed the version offline in order that a blind individual utilizing this app gains an interface to the net on every occasion they should apprehend the notes they're carrying. This framework could be improved as to use the classification technique to examine the authentic or foreign money notes on a low end versatile smartphone for visually disabled humans and inform clients through voice notice in local language. In destiny it could be expanded to apprehend foreign foreign money.

References

[1] Noura A, Semary ,Sondos M. Fadl,Magda S. Essa,Ahmed F. Gad ," Currency recognition system for visually impaired: Egyptian banknote as a study case", 5th International Conference on Information & Communication Technology and Accessibility (ICTA), 2015.

[2] Bruce Poon, Md Ashraful Amin," Recognizing Bangladeshi Currency for Visually Impaired", Conference Paper *in* Communications in Computer and Information Science,July 2014, DOI: 10.1007/978-3-662-45652-1_14

[3] Snehal Saraf,Vrushali Sindhikar,Ankita Sonawane,Shamali Thakare," Currency Recognition System For Visually Impaired", IJARIIE-ISSN(O), vol. 3, no. 2 , pp.2395-4396, 2017.

[4] Ahmed Yousry, Mazen Selim, Mohamed Taha," Currency Recognition System for Blind people using ORB Algorithm", *International Arab Journal of e-Technology,* vol. **5**, no. 1,pp. 34-40,January 2018

[5]Mahesh Pawar, Ketan Desai, Sanjay Gupta, Trupti Chavan, Kirti Mhamunkar," Object Detection and Currency Recognition Using CNN", Cikitusi *Journal For Multidisciplinary Research.,*Vol. **6,** no. 4, pp.357-362,April 2019

[6] Ali Abd Almisreb,Mohd.A. Saleh," Transfer Learning Utilization for Banknote Recognition: a Comparative Study Based on Bosnian Currency", Southeast Europe Journal of Soft Computing,vol.. 8 no.1,pp.28-31,March 2018.

[7] Srivastava, N.,Geoffrey Hinton, A. K., Sutskever,I.,Salakhutdinov, R.Dropout,"A simple way to prevent neural networks from overfitting", *Journal of Machine Learning Research,*Vol. **15**, pp. 1929–1958,2014.

[8] Tong, C., Lian, Y., Qi, J.,Xie, Z," A novel classification algorithm for new and used banknotes. Mobile Networks and Applications, , vol **22**, no 3,pp. 395–404,2017.

[9] V. D. Hoang and H. T. Vo, "Hybrid discriminative models for banknote recognition and anti-counterfeit,"NICS 2018 - Proceedings of 2018 5th NAFOSTED Conference on Information and Computer Science, 2019.

[10] V. Abburu, S. Gupta, S. R. Rimitha, M. Mulimani, and S. G. Koolagudi, "Currency recognition system using image processing," 10th International Conference on Contemporary Computing, IC3 2017, 2018.

[11] Y. Liu, J. He, and M. Li, "New and Old Banknote Recognition Based on Convolutional Neural Network," ICoMS 2018: Proceedings of the 2018 International Conference on Mathematics and Statistics, pp. 92–97, July 2018.

[12]M. Jiao, J. He, and B. Zhang, "Folding Paper Currency Recognition and Research Based on Convolution Neural Network," International Conference on Advances in Computing, Communications and Informatics, ICACCI 2018, 2018.

[13] H. C. Shin et al., "Deep Convolutional Neural Networks for Computer-Aided Detection: CNN Architectures, Dataset Characteristics and Transfer Learning," IEEE Trans. Med. Imaging, 2016.

[14] T. D. Pham, D. E. Lee, and K. R. Park, "Multinational banknote classification based on visible light line sensor and convolutional neural network," Sensors (Switzerland), vol. **17,** no. 7, 2017.

[15] P.Selvi Rajendran,"Virtual Bulletin Board using Man-Machine Interface (MMI) for Authorized Users", Indian Journal of Science and Technology, vol.**12**,no.34, September 2019.

[16] P.Selvi Rajendran ,"AREDAI: Augmented Reality based Educational Artificial Intelligence System, International Journal of Recent Technology and Engineering (IJRTE)", vol.**8** ,no 1, May 2019.

[17] P.Selvi Rajendran,"Virtual Information Kiosk Using Augmented Reality for Easy Shopping",International Journal of Pure and Applied Mathematics (IJPAM),Vol. **118,** no. 20, pp. 985-994, 2018.

[18] H. Hassanpour, Payam Masoumifarahabadi," Using Hidden Markov Models for paper currency recognition", Expert Systems with Applications,vol. **36**,no.6,pp.10105-10111, August 2009.
[19] Manish Shankar Kaushik, Aditya Bihar Kandali," Recognition of facial expressions extracting salient features using local binary patterns and histogram of oriented gradients", International Conference on Energy, Communication, Data Analytics and Soft Computing (ICECDS), 2017.

AUTHOR INDEX